CAROLINAS
FRUIT & VEGETABLE GARDENING

How to Plant, Grow, and
Harvest the Best Edibles

Quarto is the authority on a wide range of topics.

Quarto educates, entertains and enriches the lives of our readers—enthusiasts and lovers of hands-on living.

www.quartoknows.com

First published in 2013 by Cool Springs Press, an imprint of Quarto Publishing Group USA Inc., 400 First Avenue North, Suite 400, Minneapolis, MN 55401 USA
Telephone: (612) 344-8100 Fax: (612) 344-8692

quartoknows.com
Visit our blogs at quartoknows.com

Cool Springs Press titles are also available at discounts in bulk quantity for industrial or sales-promotional use. For details write to Special Sales Manager at Quarto Publishing Group USA Inc., 400 First Avenue North, Suite 400, Minneapolis, MN 55401 USA.

ISBN-13: 978-1-59186-563-6

Library of Congress Cataloging-in-Publication Data

Elzer-Peters, Katie.
  Carolinas fruit & vegetable gardening : how to plant, grow, and harvest the best edibles / Katie Elzer-Peters.
     p. cm.
  Other title: Carolinas fruit and vegetable gardening
  Includes index.
  ISBN 978-1-59186-563-6 (softcover)
  1. Gardening--North Carolina. 2. Gardening--South Carolina. 3. Fruit--North Carolina. 4. Fruit--South Carolina. 5. Vegetable gardening--North Carolina. 6. Vegetable gardening--South Carolina. I. Title. II. Title: Carolinas fruit and vegetable gardening.

  SB453.2.N8E49 2013
  635.09756--dc23

                        2013013206

Acquisitions Editor: Mark Johanson
Design Manager: Cindy Samargia Laun
Layout: S. E. Anderson

Printed in China

10 9 8 7 6 5 4

CAROLINAS
FRUIT & VEGETABLE GARDENING

How to Plant, Grow, and
Harvest the Best Edibles

KATIE ELZER-PETERS

COOL
SPRINGS
PRESS
Home and Garden Experts™

MINNEAPOLIS, MINNESOTA

CONTENTS

PART II

DEDICATION

For Margaret Shelton, the greenest thumb in North Carolina.

ACKNOWLEDGMENTS

I always liked to write, but without the patient instruction of Mrs. Wilhoite, Mrs. Quandt, Mrs. Libby, Dr. Swasey, Dr. Garrison, and a graduate student teacher at Purdue whose name I've long forgotten, but whose influence was enormous, I would not be the same person today, and would not have the privilege to do what I do. Billie Brownell has continued to nurture my writing over the past few years, and I appreciate all of the insights and improvements she has helped me make. Billie, I hear your voice in my ear when I write—and that's a good thing.

My thanks go to Roger Waynick, original Publisher of Cool Springs Press, and to Mark Johanson, my Acquiring Editor for this project, each for taking a chance on me.

Once you start working on a book, there are people along the way that take the word doc and make it into a beautiful book. For that I have to thank the copy editors and designers and hort editors and indexers for their work. Without them, it's all just words on the computer that might or might not make sense, and are certainly not very interesting to look at.

Every writer needs a cheerleader, and Tracy Stanley at Cool Springs has been that and more. She is the most patient, encouraging, and helpful editor one could ask for. This has been much easier with you to help me. Thank you so much, Tracy!

The Owner and Chef of Epic Food Co., James Bain, along with his staff, have been so friendly and nice, allowing me to camp out, drink their tea, eat their chocolate chip cookies, steal their WiFi, and devour their glorious vegetable creations while working on this book. Thank you.

And I couldn't do anything without the love and support of my parents, who never said no when I wanted them to buy me a book and my husband—chief garden-waterer and dog wrangler in the house. Joe, you have the patience of a saint, and I'm glad you're mine.

PREFACE

I have been gardening since I could walk, and I have enjoyed, throughout my professional career, teaching others how to garden.

When faced with the task of making vegetable gardening an easily digestible (pun intended) topic for new gardeners, I tried to come up with a method of organization that differed from the usual A–Z list of plants.

Organization by Function, Not Name

Gardens don't grow well when alphabetized. I know this because I've tried to plant alphabet vegetable gardens for kids. The cucumbers end up growing all over the beets, and the grapes shade the eggplants. It's messy. So why should your gardening book be organized that way? Sure, it's easy to look things up alphabetically, but what if the book, through the way it is organized, could help you plan your garden?

Now we're talking! (There's always the index and the alpha-order plant list to help you look up the location of your favorite veggie's information.)

In the Carolinas, you can garden year-round. Greens and root crops flourish in the winter, and tropical vegetables such as tomatoes and peppers thrive during the hot summer. This book has sections covering cool-season gardening and warm-season gardening.

Your fruit or vegetable garden doesn't need to be this big in order for you to enjoy fresh produce!

A few holes in the leaves of this eggplant plant won't affect the fruit.

It isn't just weather you have to be concerned about, though. The plant families (stone fruits, cabbage relatives) and their growth habits (tree, shrub, vine) also influence where you plant them and how you grow them. The fruits section is organized by growth habit to make it easy for you to figure out where to plant the plants, so that they produce well for you without taking over your entire garden.

Hopefully this admittedly unusual style will help when you're ready to put shovel to soil. Just open to the section you want to tackle and go!

Never Too Ugly to Eat

If there's one piece of advice I could give to any new gardener—the one thing you can't ignore—it is this: your garden doesn't have to be gorgeous to produce well. Your tomato plant could look like it is on its last legs, but if it is still flowering and pumping out tomatoes, don't sweat it.

The eggplant leaves might have spots. The eggplants will probably be fine. Just because you see a few holes here and there doesn't mean you need to bring out the big guns and spray everything.

If you see aphids on the plants, get out the hose and spray them. If that doesn't work, try insecticidal soap. (Read on to learn how to deal with aphids and other pests.) If you see a giant tomato hornworm, pluck it and squash it. But don't feel discouraged if your garden doesn't look like a magazine cover. Is what you're growing tasty? Good. That's what you're going for.

Now get out there and get your hands dirty! Get the book dirty, too! That's what it's for.

—Katie

GROWING EDIBLES IN THE CAROLINAS

I firmly believe that if you understand how plants grow, you'll take better care of them and have better results. It isn't enough to know just the how-to. To be successful

at gardening, you also need to know the why-do. With a bit of information, gardening suddenly goes from mysterious to predictable. You'll learn that if you prune a plant this way it will grow that way.

Part 1 of this book gives you everything you need to know to grow the best edibles in your home garden.

Carrots are easy to grow in the Carolinas.

Scarlet runner beans growing up trellises are as pretty as they are tasty!

You'll learn about:

Gardening within the Carolinas—regional differences in soil, rainfall, temperature, and so on

Seasonal gardening—what it means and how to make the most of each season

Garden planning—getting the most out of your space, large or small

Soil—the most important part of every garden, and the trickiest aspect of Carolina gardening

Selecting plants—varieties for our area, whether to buy transplants or seeds, and how to plant them

Growing and maintaining the garden—how to water, when to fertilize, what to use for mulch, how to deal with pests, and so forth

USDA PLANT HARDINESS ZONE MAPS

North Carolina and South Carolina

The USDA Plant Hardiness Zone Map indicates the average minimum cold temperatures for regions. The map was reissued in 2012 and reflects a warming trend affecting gardeners.

All plant labels and catalogs list hardiness zones to aid in plant selection—particularly trees, shrubs, and perennial herbs and vegetables. A plant is considered "hardy" to a zone if it can withstand the average minimum temperature in that area. If your area is too cold for a plant to live through winter, you can still grow it as an annual. Hardiness zones are not as applicable to annual vegetables, such as tomatoes or peppers, that you grow for only one season.

Sow pea seeds for a fall cover crop.

There are other temperature-related factors affecting plant growth other than annual minimum temperature, including:

Chilling hours: hours below 45 degrees F, mostly related to fruit production
Heat: extreme summer heat that causes blossom drop in vegetables
Temperature fluctuation: can cause bolting and flowering in cool-season crops

These factors are explained later in the chapter. Because all plant labels indicate plant hardiness, they are a good place to start in learning how to pick out the right plants.

Look at the hardiness zone map and see which zone corresponds to your area. Remember both the zone number and the temperature range—they will come in handy when selecting plants.

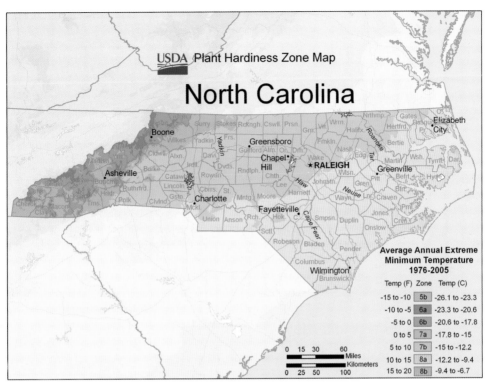

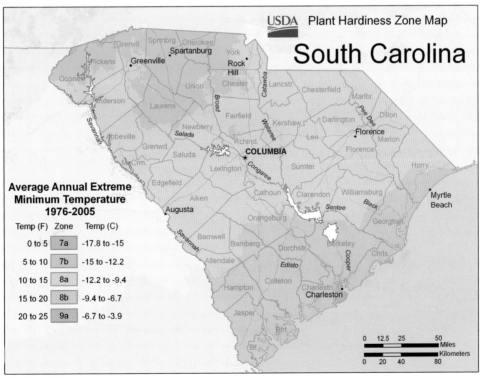

USDA Plant Hardiness Zone Map, 2012. Agricultural Research Service, U.S. Department of Agriculture. Accessed from http://planthardiness.ars.usda.gov.

CHAPTER 1

THE CAROLINAS REGION

Gardeners in the Carolinas experience long, hot, humid summers and cool winters within USDA hardiness zones 6 through 9. Mountainous western North Carolina is the coolest, with zone 6a and 6b designations, while coastal South Carolina is the warmest, with large swaths of zone 9a reaching from the coast to the edge of the Piedmont region in the center of the state. Weather isn't the only factor influencing gardening

The gardening conditions that affect you and your plants, from soil characteristics to weather fluctuations, are directly related to where you live.

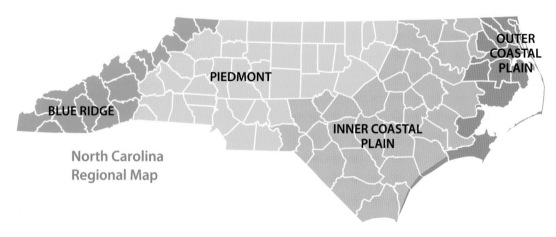

OUTER COASTAL PLAIN

PIEDMONT

BLUE RIDGE

INNER COASTAL PLAIN

North Carolina
Regional Map

South Carolina
Regional Map

Coastal Region

The coastal region of the Carolinas extends from the coastline to the interiors of both North and South Carolina. While only gardens just off the ocean experience issues with salt spray, all of the gardens within a couple hundred miles of the coast benefit from the temperature-moderating effects of the ocean. The North Carolina coastal region extends from the coast inland to Raleigh and Fayetteville. In South Carolina, coastal characteristics are found from Charleston inland to Columbia.

High summer heat and humidity paired with cool, moderate winter temperatures are the most marked weather conditions in coastal regions. Soils in these regions are typically loose, sandy, fast draining, and low in organic matter, although some areas have poorly drained clay and muck soils.

Hardiness zones: Coastal North Carolina: 8a, 8b; Coastal South Carolina: 8b, 9a

Soil characteristics: Sandy, fast draining, low in organic matter

Unique issues: Salt spray, summer heat and humidity

Piedmont Region

The term *piedmont* means "foot of the mountain." The Piedmont region of the Carolinas covers much of west-central North Carolina from Raleigh, continuing southwest to Charlotte, and dipping down to Greenville, South Carolina, encompassing the very uppermost left corner of that state. Soils in the Piedmont region have a much higher clay content than soils in the coastal regions, leading to drainage problems. Summers are hot and humid, often with stagnant air. Winters are cool and mild, with the occasional ice storm that can be damaging, particularly to fruit trees.

Hardiness zones: 7a, 7b, 8a
Soil characteristics: Higher in clay than soils on the coast; can be hard and difficult to work, with some drainage problems
Unique issues: Ice storms

Mountainous Region

Western North Carolina from Boone southwest to Asheville is the mountainous region of the Carolinas. This region is the coolest during winter months but still experiences hot, humid summers. Soils in the mountains are rocky but have more organic matter than soils at the coast. Peach and apple orchards are common sights in the mountains, and this area is perfect for fruit tree cultivation.

Hardiness zones: 6a, 6b
Soil characteristics: Rocky, mineral rich, acidic
Unique issues: Snow, colder temperatures

How to Use Frost Tables

Seed packets and plant labels usually have language that refers to the average last-frost date or the average first-frost date, such as:

- Plant inside six to eight weeks before the average last frost. Transplant outside after danger of frost has passed.
- Sow seeds directly outside twelve weeks before the last frost.

Frost tables refer more to the length of different portions of the gardening season than to the severity of the winter, as the USDA hardiness zones do. Use frost tables to judge when to plant cool-season crops to take advantage of cool weather and to plan for warm-season crops during the time of the year they grow best.

Frost Tables by City

NORTH CAROLINA City	Last Probable Spring Freeze	First Probable Fall Freeze	Probable Number of Frost-Free Days
Asheville	May 18	Oct. 7	157
Charlotte	May 12	Oct. 28	199
Elizabeth City	Apr. 26	Oct. 29	200
Elizabethtown	May 1	Oct. 23	194
Fayetteville	Apr. 30	Oct. 23	194
Franklin	May 29	Oct. 3	146
Goldsboro	Apr. 25	Oct. 28	203
Greensboro	May 7	Oct. 18	178
Greenville	Apr. 25	Oct. 23	193
Hendersonville	May 14	Oct. 10	163
Hickory	May 6	Oct. 17	179
Mount Airy	May 15	Oct. 9	157
New Bern	Apr. 18	Nov. 1	211
North Wilkesboro	May 19	Oct. 6	156
Raleigh	May 7	Oct. 22	185
Tryon	May 6	Oct. 19	179
Wilmington	Apr. 28	Oct. 24	194

SOUTH CAROLINA City	Last Probable Spring Freeze	First Probable Fall Freeze	Probable Number of Frost-Free Days
Aiken	May 5	Oct. 17	183
Anderson	May 4	Oct. 19	183
Beaufort	Apr. 9	Nov. 19	246
Camden	May 1	Oct. 23	188
Charleston City	Mar. 28	Dec. 5	275
Columbia	May 1	Oct. 23	190
Conway	Apr. 16	Nov. 6	219
Darlington	Apr. 26	Oct. 23	196
Edisto Island	Apr. 13	Nov. 16	236
Florence	Apr. 23	Oct. 29	208
Georgetown	Apr. 27	Nov. 8	218
Greenville-Spartanburg	May 6	Oct. 22	186
Greenwood	May 8	Oct. 20	180
Kingstree	May 1	Oct. 23	191
Saluda	May 4	Oct. 19	182
Summerville	Apr. 29	Oct. 29	203
Sumter	Apr. 22	Oct. 25	201

CHAPTER 2

SEASONAL GARDENING

Gardening seasons in the Carolinas are referred to as "cool season" and "warm season." These two seasons are so important that this entire book is organized around them. For the most part, "cool" and "warm" are apt descriptors for the conditions that certain plants like. The two seasons have another defining factor beyond temperature, though, and that is light. There is less daylight and a longer night during the cool season. There is more daylight and a shorter night during the warm season.

It just so happens that one of the cool-season plants everyone enjoys eating, spinach, is also a long-day plant that flowers when the days are long and the nights are short. That's good for gardeners because we don't want spinach to flower when we're growing it to eat. We just want lots of tender leaves. It's funny how that works!

For the most part, you don't have to worry about day length in the vegetable garden. However, if you find yourself having difficulty with a plant, and you're sure that you've gotten everything else right, look it up to see if day length is a factor in the way it grows. You might be beating your head against the wall for no reason.

BIENNIAL VEGETABLES

Some plants defy neat categorizing—they grow equally well in warm weather and cold weather. These plants are considered to be biennial. They won't flower and set seed until their second year. Swiss chard and arugula are the two main traditional "cool weather" vegetables that will grow just fine

Swiss chard is a biennial vegetable.

throughout summer if they are planted in spring. Kale will grow throughout summer, as well, though the leaves can be tough and bitter during hotter weather.

Cold and heat both influence plants. This concept can be confusing to gardeners, because most plant labeling covers only USDA hardiness zones, and those zones relate to cold and not heat. Zones are useful for only one temperature-related bit of information: whether the plant can withstand the coldest air temperatures likely to occur in a region.

There are many other temperature-related factors, beyond cold-hardiness, that affect plant growth, including soil temperature, air temperatures (beyond the minimum temperature), and fluctuation in temperatures. Temperature affects when you can plant, when plants will flower and fruit, and when you need to harvest fruits and vegetables.

Cool-Season and Warm-Season Gardening

In the Carolinas, there are two growing seasons: warm and cool. The cool season runs from about October or November through April or May (depending on where you garden). The warm season runs from May or June through September or October. If you plan your Carolina garden around no other guiding principle than this, you will be well in front of people who don't. The simple fact is:

- Cool-season herbs and vegetables will not grow during summer. You will frustrate yourself endlessly if you try to make them do so. The same goes for trying to grow warm-season vegetables and herbs in cooler weather.
- Just because the stores stock tomato transplants in March doesn't mean you should plant them outside in March.

Most of the edible plants we grow came from somewhere else besides North America. Blueberries are native, as are cranberries. Their history provides clues to the conditions in which they grow best:

- Cabbage, and cabbage-family plants (brassicas) were domesticated as garden plants in cooler, northern regions of Europe. Cabbages enjoy long, cool, but not cold, winters in the Carolinas.
- Tomatoes are native to tropical highlands and are long, perennial vines in their natural habitat. We grow them as summer annuals during the warm season of our more temperate climate. Once a frost hits, they're done.

Bolting

When cool-season vegetables, most of which are true "vegetables" (we eat the vegetative parts of the plants—leaves, roots, stems), move from the vegetative stage of growth to the flowering stage of growth—an action called bolting—it's time to stop eating them. All of a sudden, from the nice little clump of lettuce leaves you've been picking, a long stem starts to grow. Then flowers appear. The plant has bolted, and it is completing its life cycle, ending in flowering and seed production.

Lettuces, cabbages, broccoli, cauliflower, turnips, and radishes are all vegetables that will bolt. Dill and cilantro are two cool-season herbs that bolt. (Unlike lettuce seeds, though, dill seeds and cilantro seeds (coriander) are useful in the kitchen.) In the Carolinas, you run into bolting during two times of the year: fall and spring.

Fall Bolting

Fall weather in our area can include rapid temperature changes from one day to the next. If you plant lettuce, mesclun greens, or mustards when it is cool outside and then experience a few weeks of warm weather followed

by another cool-down, you could notice the plants starting to bolt. If your plants do this, you will need to replant after the weather settles down into a more consistently cool pattern. Occasionally cabbage and cauliflower will be affected by these fall temperature swings, as well.

Spring Bolting

Any cool-weather plants that you planted in fall or early spring have the potential to bolt in late spring as the temperatures warm up and the days get longer. Resist the urge to keep your fall-planted cabbages in the garden for too long in spring, because they will start to crack and

Collard plants bolting (flowering and producing seeds).

Broccoli plants bolting. The part of the broccoli that we eat is the flower. Once the broccoli heads start to grow and bloom, the window of harvest is over.

flower. Pick broccoli heads when they are still tight. The part of broccoli that we eat is actually the plant's undeveloped flower head. If you let the heads sit in the garden, they will start to stretch out and bloom.

Soil Temperature

You can buy an inexpensive soil thermometer to help you gauge when to plant different edibles. Cool-season vegetables germinate well when the soil is at least 50 degrees F. Most will sprout when soil temperatures are 50 to 70 degrees F. Warm-season vegetables grow best when soil temperatures are at least 65 degrees F. You can plant these when soil temperatures are lower, but the plants won't grow.

Diseases are also more of a problem in cool, wet soils. Sweet corn seeds will rot in the ground if planted too early, while the soil is still cold. Damping off, pythium, and root rot diseases can strike seeds that are planted too early in the year, when the soil hasn't warmed up yet.

Soil thermometer

Tropical Annual Edibles Like It Hot

These favorite summer vegetables are native to tropical areas and grow only when soil temperatures are at least 65 degrees F. Don't plant these too early:

Bean	Pepper	Tomatillo
Corn	Pumpkin	Tomato
Cucumber	Summer squash	Watermelon
Eggplant	Sweet potato	

Biological Indicators

If you read older gardening books, you'll sometimes see directions with biological indicators rather than dates. There might be instructions to plant a certain vegetable "when the redbuds are in bloom." Rather than using dates as an indicator of when to plant, these instructions use what's going on in the world around the garden as information for timing of planting. You can learn your own indicators. Observe what's happening in and around the vegetable garden when you plant certain plants. Over the years, you'll learn that you will sow your last crop of lettuce when the azaleas are blooming, and you'll harvest your first strawberries when the daylilies start flowering.

Air Temperature

Cold air can be shocking to new transplants. Extremely hot air (85 degrees F and hotter) will cause flowers to drop off tomatoes, reducing the eventual tomato harvest. Some plants (many fruit trees) need a certain number of hours below 45 degrees F in order to produce fruit. Beyond a killing freeze, here's what you need to know about air temperature and plant growth.

Chilling Hours

Some plants not only like, but *need* colder air. Most fruit trees need a certain number of hours with air temperatures below 45 degrees F, called chilling hours. Plants with chilling hour requirements must meet the minimum number of chilling hours in order to break dormancy and bloom. (Remember: no blooms, no fruits!) During warmer winters, fruit trees can get physiologically "confused," blooming early, late, or not at all.

In the Carolinas, it's important to select fruit tree varieties with relatively low chilling hour requirements. Gardeners in the mountains of western North Carolina will experience more chilling, but the whole region gets less chilling than areas of the Midwest, Pacific Northwest, or Northeast.

Before buying fruits, look at the number of chilling hours your area gets.

Minimum Chilling Hours Available by Region
Mountains: more than 1,200
Foothills and Piedmont: 800–1,000
Coastal regions (northern): 600–800
Coastal regions (southern): 400–600

Corn plants roll up their leaves to prevent water loss.

WHAT'S WRONG WITH MY CORN?

When corn plants don't have enough water available to take up (the soil is too dry), they roll their leaves up to reduce the amount of surface area exposed, thus lowering the amount of water lost through the leaves. If the corn leaves are rolled up and "pointy," you need to water the plants!

Heat and Fruit Set

Heat, the other side of the thermometer, affects fruit set in a favorite garden crop: tomatoes. When daytime temperatures rise above 85 degrees F, tomato flowers fall off the plants without being pollinated. No flowers means no fruit.

Temperature Variations within the Carolinas

Elevation, latitude, and proximity to the coast (and the moderating effects of the ocean) all contribute to temperature differences between areas within the Carolinas. There can be a month or more between the optimum tomato-planting time in Asheville, North Carolina, and that in Charleston, South Carolina. Do your homework before deciding what to plant and when.

CHAPTER 3

PLANNING YOUR GARDEN

Plan before you plant. It's easier to imagine everything on paper than it is to relocate it once you've put shovel to soil.

Where to Put Vegetable Beds

Sunlight is the biggest limiting factor in where you can put your vegetable beds. Keep in mind that where the sun is during winter and during summer are two different things. For one thing, sun hits places in winter that it doesn't reach in summer, because deciduous trees have lost their leaves. The sun is also at a higher angle in the sky in summer than in winter, which means that shadows created by things such as houses or fences fall in different places.

Planting edible trees or shrubs along a lot line is a good way to use this space and a friendlier way to mark the edge of your property than a fence.

There are also other factors to consider when placing the vegetable garden:

Your sightline and the traffic patterns around your house: If your vegetable beds are stuck in a back corner where you don't have to view them or walk by them regularly, you'll pay a lot less attention to them, which is not good for your plants.

Proximity to a hose hookup: It is (literally) a drag to have to run 100 feet of hose across sidewalks and through a garden gate to get to the vegetable garden. (Trust me, I've done it. I do it. I have no other choice because of where the sun hits my yard.)

Location of garden tools: The closer you get the bed to the tools, the easier it will be to do maintenance.

Take it from someone who has had to break all three of these rules to get a vegetable garden in her yard: if you can, follow the above tips.

To Fence or Not to Fence?

Whether or not you should fence your vegetable garden is a matter of preference and necessity. If you have deer problems, absolutely look into some kind of fencing—but it had better be tall! If you don't have critter problems (two-legged or four-legged), a fence can be more of a pain than a help. For one thing, it can block sunlight. Fencing can also be expensive. I side with the "not to fence" crowd, unless you can't grow vegetables without one.

Electric Fence for Vegetable Gardens

If you have bad pressure from deer in the area, you can put up an electric fence that's relatively unobtrusive. Fiberglass rods and a single electrified strand will do it. The trick is, you have to bait the fence by clipping foil-wrapped peanut butter to the electrified wire. The only way to train the deer to stay away from a single-strand fence is to make them put their tongue on it. All it takes is one time—then they'll stay away.

Siting Fruit Trees and Shrubs

I like to plant my fruit trees and bramble shrubs along the edges of my property. Primarily, that's because it's the only place I have room. If you're planning to plant fruit trees and shrubs in an older subdivision, you might find that the only areas you have to work with, without having to uproot a bunch of other plants, are the edges of your property. That's fine! It's nicer (and less expensive) to put up a "fence" of blackberries than an actual fence. Once fruit trees start to bear, they can also be messy. (It's impossible to get all of the fruit from the tree into the harvest basket—there will be some that ends up on the ground.) Keeping the trees in one area makes cleanup of the rest of your yard easier.

Sage varieties line a short pathway outside the kitchen door. By planting herbs close to the kitchen, you'll be sure to use them more.

Incorporating Edibles into the Landscape

Some edibles are easier to plant in the landscape than others. If you aren't ripping out your entire front yard to turn it into a vegetable garden, but you have limited space, you can plant certain edibles as foundation plantings or in the flower beds. Overall it is a lot easier to care for vegetables and fruits if they're in one location, but if you have the opportunity to redo your landscaping, or you want to shoehorn in some extra plants, these are some edibles that play well with others:

Blueberries: In addition to their tasty fruits, these deciduous shrubs have beautiful red fall color.

Apricots and almonds: These bloomers are beautiful landscape specimens.

Swiss chard, kale, and cabbage: These cool-season greens are as beautiful as they are tasty. As an added bonus, chard and kale will grow in partial sun—a rarity for most edibles.

Sorrel, fennel, dill, and parsley: All four of these herbs will grow in full sun to partial shade, and they look quite at home in a perennial border.

Culinary sage: Instead of growing Mexican sage, plant culinary sage instead. You'll get the same gray-green leaf effect, but the culinary sage is edible.

Creeping thyme: Anywhere you need a ground cover in full sun, you can plant creeping thyme. If you happen to step on it, you'll release some of the fragrance of the leaves.

Placing Perennial Edibles

You might decide that you want to grow some perennial edibles in your vegetable garden. You can put these vegetables and fruits wherever you

---

GROWING TIPS

Don't forget the pollinator trees. Not all fruit trees need a pollinator tree, but many do. For each fruit type, select a second variety that blooms around the same time as your other choices, and plant it in the same row as the others.

want, but it is a good idea to put them somewhere a bit out of the way, where you won't mind them staying put. They're all different, so there's not a "one plant fits all" solution.

Asparagus beds take up a lot of room, and the plants grow to be quite tall during summer. Globe artichokes are large plants that also get to be tall, so you could site these on the north side of the vegetable garden.

Sorrel is a shorter, decorative herb. It would be at home in the perennial garden or as a vegetable bed border plant.

Perennial Edibles

Asparagus	Rhubarb
Globe artichoke	Sorrel
Radicchio	Strawberry

The Edible Container Garden

There are entire books about edible container gardening filled with more than we can cover here. If space is limited though, or you want to grow plants that spread (mint) or are marginally hardy (citrus), containers are the way to go. If your soil is infected with diseases or nematodes, containers may be your only option.

A lettuce bowl is a fun way to grow salad greens if you're limited on time and space.

Lettuce bowls: If you can't bring yourself to do a full-scale seasonal switch of your vegetable garden, planting a lettuce bowl or two is a good alternative. In two 12- or 16-inch-diameter bowls, you can grow enough lettuce to pick four to six salads a week.

Marauding mint: Mint will take over an entire yard in one season if allowed. Want to have fresh garnishes for your cocktails? Plant just one pot of mint—you'll have plenty.

Broccoli and cole crops: Plants in the cabbage family have to be rotated to different areas of the garden each year for three years. If you have raised beds, this can be difficult. Instead of planting these plants in the ground, fill a few whisky barrels with new soil each year and plant your broccoli in there to avoid soilborne diseases.

Keep mint from taking over the garden by growing it in a pot.

Raised-Bed Gardening

If you're interested in growing vegetables, you've probably heard about raised-bed gardening. So what's the fuss all about? Raised-bed gardening makes growing vegetables much easier for people with limited space or time, mobility challenges, or disease-infested soils. By gardening in raised beds, you can grow more food in less space and more easily control the soil and growing conditions, while contending with fewer weed problems.

Assembling a Raised-Bed Kit

2 You can fill the beds with soil or compost—no weed cloth needed. If you have Bermuda or zoysia grass, you will need to kill the grass before adding soil. If the soil underneath your bed is clay or compacted, loosen it by tilling or digging before filling beds.

1 Raised-bed kits are easy to assemble.

3 Raised beds are easy to reach and plant.

4 You can go from zero to garden in just an hour!

Benefits of Raised-Bed Gardening

Easy to assemble

Elevated gardening area

Less soil compaction

Soil warms up faster

Can control the quality of the soil

Can grow more in less space

Fewer weed problems

Challenges of Raised-Bed Gardening

High startup costs

Limited space for growing

Difficult to move or disassemble

Build Your Own or Buy a Kit?

You can make your own raised-bed garden out of stacked concrete blocks, wood timbers, discarded bricks, or landscape timbers. Just about the only material you should not use to make a raised-bed garden is railroad ties. Those are treated with so many chemicals that you'd be working against yourself from the get-go.

Myth or fact? You shouldn't use treated wood to build raised-bed gardens.

Myth. The effects of treated wood extend only a few millimeters from the surface of the wood itself. Using untreated wood is a sure way to make sure you have to rebuild within a few years. If you have access to cedar planks and can afford them … you can have the best of both worlds—a wood garden bed frame with natural anti-rotting properties.

If you're not handy with a saw and a hammer, check out the many raised-bed kits on the market. Most of them are so easy to assemble that you can go from zero to planted vegetable garden bed in under an hour.

Raised-Bed Maintenance

There are two types of maintenance involved in raised-bed gardening: maintaining the bed structure and maintaining the bed soil. It's a good idea to walk around the raised beds at the beginning of each year to make sure that everything is in its place. Do any nails need to be pounded in? Is there dry rot in one of the boards? Are any pieces of plastic cracked?

Because raised beds are gardened more intensely than landscape beds or row gardens, producing more food per square foot, you also need to maintain the soil. When you change over a raised bed from cool-season to warm-season vegetables and back again, add some compost to each of the beds and incorporate it into the beds with a garden claw. You can purchase compost or make your own. Worm castings are also a good soil additive.

Garden Planning for Raised Beds

If your entire vegetable garden is made of raised beds, it's tempting to keep them all planted so that everything looks nice and neat. Resist the urge! The beauty of raised-bed gardens is that it is easy to move from season to season, and crop to crop, without having to get out a big rototiller and turn over a half-acre garden. Give yourself flexibility by leaving areas open. If you have six raised beds, plant two in peas in early spring. Leave two for successive plantings of lettuce and cool-season herbs. Plant one with cabbage in fall to harvest in spring; leave one fallow so it is ready for the first plantings of beans or tomatoes while the rest of the cool-season crops are still finishing up.

CHAPTER 4

BUILDING GREAT SOIL

Have you heard the phrase, "plant a five-dollar plant in a twenty-five-dollar hole"? It's true. The soil is where plants take up oxygen, water, and nutrients for the processes they need in order to grow. The soil holds and releases nutrients for the plants. Plants draw water from the soil. Characteristics of the soil make it easy or difficult for plant roots to grow, spread, and reach more water, nutrients, and oxygen. Another phrase you see often in gardening literature is, "feed the soil, not the plants." Time spent getting the soil where it needs to be in order to support healthy plant growth is time saved trying to fix problems later.

A compost bin can keep roots out of your compost and help contain the pile.

Fertilizer, whether synthetic or organic, is almost always necessary in the Carolinas.

What You're Starting with: Carolina Soils

Soils in the Carolinas are not rich in the organic matter that facilitates good water and nutrient-holding capacity. The glacial movement that deposited the nutrient-rich, deep, silty loam soil of the Midwestern farmland did not touch the Carolinas. Our soils are typically sandy to rocky, fast draining, and low in organic matter. The exception to this is the Piedmont, where there are large swaths of clay. It is easy for plant roots to grow into the soil, but they might not find much there.

As in the rest of the eastern United States, the pH for most soils in the Carolinas tends toward the acidic side, between 5.0 and 6.5, but is often higher at the coast. Some edible plants are happy with this, while others need a higher pH.

So while the soil you're starting with probably has some good things going for it, you'll want to improve it to increase your harvest. Here's what to do.

Add Organic Matter

Some people would tell you to test the soil before you do anything else. You can do that, but once you add compost or additional organic matter, such as worm castings or blood meal, you will have changed the composition—and possibly the pH—of the soil, which makes your soil test useless.

Why would you add more compost to the soil before testing it? You can *always* add compost and will always need to. In our warm, moist climate, compost breaks down quickly. After you garden in the Carolinas for a few years, even if you garden in raised beds, you will find that, no matter how much compost you add, the sand, rocks, and everything else you're trying to overcome by adding the compost continues to move its way up into the garden soil.

How to Compost

You can buy compost if you're short on time. You can (and should) also make your own. Composting is a great way to cut down on the amount of trash you throw away. It's also a shame to have to buy something you can make yourself from food scraps, leaves, and yard debris that you otherwise have to pay someone else to take away!

It's important to continually add compost to your raised beds and vegetable gardens.

Compost Ingredients

There are two types of materials to add to compost piles: green materials, which are high in nitrogen, and brown materials, which are high in carbon. While there isn't a precise ratio of green to brown materials required in compost piles, if you add equal amounts of each over time, you'll get usable compost faster.

Kitchen scraps

Shredded newspaper

Green Materials

Grass clippings   Kitchen scraps
Weeds   Eggshells
Green leaves

Brown Materials

Shredded newspapers   Paper towels
Dried leaves   Twigs
Sawdust   Tea bags
Wood chips   Wheat straw
Paper bags   Coffee grounds

You can compost almost any natural material, but if you're composting at home, don't put any animal products other than eggshells in the compost pile. No meat, cheese, dairy, or lasagna leftovers.

Grass clippings

Building the Pile

Build a compost pile by layering green materials and brown materials like you'd make a lasagna. Start with chopped-up dried leaves (you can run over them with the lawn mower). Then add grass clippings or kitchen scraps, and keep layering. The smaller the pieces you add to the pile, the faster they will decompose.

If you do not have enough material to begin with, you can build a compost pile over several weeks. While building the pile, continuously add material—shredded newspapers, the stems of broccoli, last summer's dead annual flowers that you pulled up, even the Halloween pumpkin. When the pile is large enough, stop adding new material. Use new material to start a new pile.

Keep the pile "cooking" (a healthy pile will heat up) by using a pitchfork or garden fork to turn the pile, mix it up and oxygenate it.

Compost pile

Compost Troubleshooting

Smelly compost piles are usually too wet. You can fix this by adding more brown materials and turning the pile so that it can dry out. Throw some chopped leaves, straw, or shredded paper on the pile.

As the items in the pile decompose, the pile will shrink. If the pile isn't shrinking, it isn't cooking. First turn the pile, and then water it. If that doesn't help—if the pile is damp but still not shrinking—add green materials to kick-start the organisms that break down the pile.

Other Soil Amendments

Adding compost to the garden is always a good idea. There are other types of organic matter that you can add to the soil to provide extra nutrients to plants, either when you're planting or when you're prepping the soil for planting.

Plant-tone

Blood Meal and Plant-Tone

Blood meal, bonemeal, and Plant-Tone are organic, slow-release fertilizers made from animal products. Sprinkle them around the plants (this is called sidedressing) and worked into the soil with a four-tine claw. Just like synthetic fertilizers, plant tones are available for acid-loving plants, vegetables, flowers, and trees. Each type has different combinations of nutrients and ingredients to best benefit the receiving plants. Use these when you're planting to add nutrients that will break down slowly over time.

Worm Castings and Soil Conditioner

Worm castings (worm poop) are expensive to buy, but they're wonderful for garden plants, particularly vegetables. Worm castings and soil conditioner are similar products, in that they are both almost entirely composed of humus, which is the most decomposed element in soil. Humus is excellent for plants. Incorporate worm castings or soil conditioner into your soil before planting and you'll be amazed at how well your plants grow. You can make your own worm castings by becoming a worm farmer. Buy a worm bin, and follow the

Worm castings

directions for procuring and feeding the worms. Worm composting is also a great way to get rid of kitchen scraps.

Garden Soil and Topsoil

Garden soil and topsoil are soils that you can use to build up areas of the garden where soil has washed away. Generally, it's helpful to mix garden soil or topsoil with compost before planting a garden bed. The compost adds nutrients and natural water-holding capacity, while the topsoil adds structure and bulk. Do not buy garden soil with synthetic fertilizer in it.

Garden soil

Manure

You can make your own compost, but you can also buy composted manure or pick it up from a farm. Cow, rabbit, chicken, and mushroom compost are the most commonly available types for purchase. If you get manure from a local farm, make sure that it has been sitting and aging for at least six months. Ammonia in fresh manure can burn your plants. Never use cat or dog droppings in the garden.

Test the Soil

After you've added compost, manure, and topsoil to create the garden bed in which you'll grow your plants, it's time to test the soil. Testing now gives you a more accurate picture of the conditions the plants will have to deal with while they're growing and any additional adjustments you need to make.

Cow manure

You can get soil test boxes and forms from your local Cooperative Extension office. You can also purchase soil test kits at garden centers and home-improvement stores so that you can test the soil at home. These kits allow you to test for the soil's pH and the presence of nitrogen (N), phosphorus (P), and potassium (K). Test the soil to see if anything is lacking before adding fertilizer, lime, or amendments other than compost to the garden.

GROWING TIP

You can pick up soil test boxes from your local Cooperative Extension office. Every test result will give you recommendations for adding nitrogen, phosphorus, or potassium and whether or not lime is needed. Collect a soil sample to submit by digging clumps of soil from different areas of the garden, mixing them together, and submitting the sample for testing.

How to Do a pH Test with a Kit

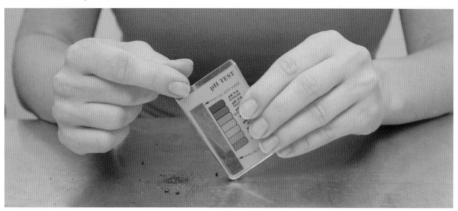

1 Add soil to the kit from the area of the garden that you want to test. Then add water according to the instructions.

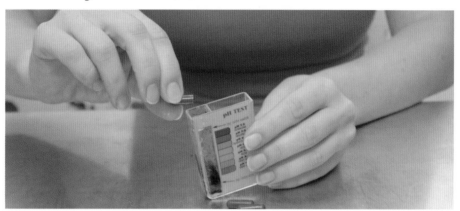

2 Add the indicator powder to the water in the container; shake the container to mix the soil, water, and indicator powder.

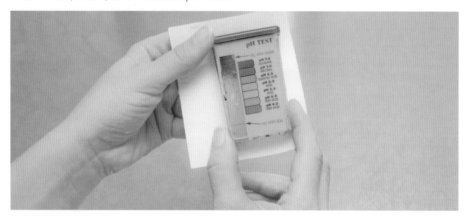

3 Hold the container against a piece of white paper so that you can check the color of the water with the soil in it against the color key on the container. The color of the water will match one of the colors in the key on the container. The reading will tell you whether the soil is acidic (low pH) or alkaline (high pH), has low nitrogen levels, or has high phosphorus levels. After testing, you'll know what to add (or not add) to the soil.

Adjust the Soil pH

The pH scale runs from 0 to 14. Most plants grow best in the range of 5.5 to 7.0, because nutrients in the soil are most available to plants at those soil pH levels. Landscape plants will usually struggle along if the soil pH is off, but edibles won't produce roots, leaves, and fruits for you to harvest if an off pH prevents them from getting the nutrients they need.

pH Ranges for Common Edibles

Asparagus	6.0–8.0	Cauliflower	5.5–7.5	Onion	5.8–7.0
Beet	6.0–7.5	Celery	5.8–7.0	Spinach	6.0–7.5
Broccoli	6.0–7.0	Cucumber	5.5–7.0	Tomato	5.5–7.5
Cabbage	6.0–7.5	Lettuce	6.0–7.0		
Carrot	5.5–7.0	Muskmelon	6.0–7.0		

How to Lower the pH

Soils with a pH below 5.5 are considered to be acidic. Soils with a pH above 7.0 are considered to be alkaline. To lower the pH of alkaline soils so that the pH is in the neutral range of 5.5 to 7.0, add aluminum sulfate according to package instructions. You can buy this at garden centers and home-improvement stores in the aisle with fertilizers and lawn care products. Some vegetables grow well with a lower pH, but most are happier with a neutral pH. Blueberries thrive in soils with a lower pH.

How to Raise the pH

Spinach and plants in the cabbage family grow best in soils with a higher pH. You'll see that calcium is available to plants only at a higher pH. To raise the pH of the soil, add lime, which is also available at garden centers and home-improvement stores. Submitting a soil sample to your local Cooperative Extension office for testing is the best way to find out how much lime you need to add.

Fertilizing Plants

Fertilizing is related to soil because you put the fertilizer in or on the soil, and the plants get the nutrients from fertilizers from the soil. Before fertilizing can be effective, the soil has to have the right properties to retain the nutrients from fertilizers and make them available to plants. If you've added compost, tested the pH, and adjusted it, and your plants still need more nutrients than those in the soil (and most edibles will), you can fertilize. More specifics on fertilizing materials, timing, and instructions are covered in chapter 6.

CHAPTER 5

PLANTING YOUR GARDEN

The fun part of growing fruits and vegetables isn't testing the soil—it's picking out what to plant! But that can be the confusing part, too. Should you buy seeds or transplants? Container-grown or bare-root plants? What *is* a bare-root plant? How many of each plant should you buy? Should you get hybrids or heirlooms? Does it matter if the plants or seeds are organic? What about fruit trees? Should you buy trees grown in Oregon to plant in South Carolina? Here's what you need to know in order to select the right plants and plant them in the right places for edible garden success.

Going Shopping

Whether you're shopping for fruits or vegetables, you will have to choose between plants or seeds labeled "organic" and those that aren't. You'll choose between buying open-pollinated and hybrid selections, newer varieties and heirlooms. You can buy plants grown locally or plants shipped from far away. Before we even discuss varieties of plants, it's important to understand these broader categories and classifications and what they mean to your garden plans.

Growing Organic

The term "organic," when used as a label on something that is for sale, means that the product has been produced or grown according to a set of government standards. The simplest way to describe those standards is that they require growing without man-made chemicals. Is it better to buy organic seeds and transplants? That's a matter of preference. Plants grown organically usually have a lower impact on the environmen,t because the growers focus more on building the soil than spreading chemical fertilizers, use biological controls as opposed to synthetic (man-made) chemical controls, and work within the natural ecosystem rather than fighting it. At the end of the day, though, a big field of cabbage is not the same as a natural forest or prairie ecosystem.

The one disadvantage of purchasing organic seeds is that they are usually not treated with fungicides and thus can be susceptible to damping off and other soilborne diseases during cool, wet weather.

Open-Pollinated Varieties versus Hybrids

If you want to save seeds from year to year, you need to grow open-pollinated varieties. The words "heirloom" and "open-pollinated" are sometimes used interchangeably, but they don't necessarily mean the same thing. "Heirloom" refers to a variety that was popular before World War II. "Open-pollinated" refers to a plant that produces stable characteristics from generation to generation. Heirlooms are usually open-pollinated, because hybridization in edible plants didn't become common until the 1970s.

Seed packets will indicate whether the plant is a hybrid or a heirloom.

Hybrids (F1 generation) are plants that grow from seeds produced by crossing two specific parent plants and saving the seeds. By crossing two plants, breeders can develop specific traits such as size, color, disease resistance, flavor, and more. Seeds from fruits of hybrid plants (F2 generation) can be saved but will not produce plants identical to their parents. You sacrifice the ability to reliably save seeds for next year for other perks like not losing your plants to a disease problem. The word "hybrid" does not mean that a plant is a GMO. Hybrids are not inherently bad, unless you want to save seeds from year to year.

Locally Grown versus Ordered from Out of Town

You can order any plant you want online. There are some restrictions about shipping to certain states that limit what you can get, but those primarily apply to states with large acreages of commercial agriculture. Just because you can order whatever you want from wherever you want doesn't mean you should.

Annual vegetables that you grow for one season can come from anywhere, and you're unlikely to have problems as long as the plants have been responsibly grown and cared for. Fruit trees and shrubs are a different matter. Plants grown for several years in the cool, moderate weather of the Pacific Northwest just won't be as well adapted to the hot, humid weather of the Carolinas. When you buy fruits, ask where they were grown. Just because you're buying them at a local garden center does not mean they were locally grown. Equally important is to buy a variety adapted to our climate.

Buying the Healthiest Plants

You want to choose healthy plants that will take off and grow as soon as you plant them. You also want to purchase plants that have the best chance of staying healthy. Vegetables and fruits can have big problems with pests and diseases, but you can, in some cases, select plants that have resistance to pests or diseases. To avoid problems later, do your due diligence when picking.

Signs That a Plant Is Healthy

It's growing. Look for new leaves at the end of stems and branches.

It looks "in proportion" with its pot. A healthy plant won't have roots growing out of the bottom of the pot, and it won't be spilling out of the pot.

It has green leaves. Unless the plant's leaves are supposed to be a different color (check the picture on the plant tag), its leaves should be a nice, medium solid green, free of spots or blemishes. Vegetables that have been sitting at the garden center for a long time start to turn yellow from lack of nutrients.

Signs That a Plant Is Not Healthy

The plant stems are mushy and rotten.

The tips of the branches are dying. If a plant has a root problem, you can tell by looking at the ends of the branches. If the leaves are turning brown and dying, leave the plant at the garden center.

There are spots on the leaves that could indicate a disease problem.

The plant leaves are crispy, the edges are brown and dry, or the plant is droopy, all of which mean the plant hasn't had enough water.

The plants are overgrown—growing out of the pot, falling over, and trailing along the shelves.

If a plant looks rotten, it is rotten. Don't buy it.

Selecting for Disease Resistance

You can buy plants and seeds that have some level of resistance to common diseases affecting those types of plants. Look for these types of abbreviations on plant tags, labels, and catalog descriptions:

BCMW: Resistance to bean curly mosaic virus

CMW: Resistance to cucumber mosaic virus

Foc, Foc 1: Resistance to fusarium yellows

PM: Resistance to powdery mildew

PVY: Resistance to potato virus Y

TMV: Resistance to tobacco mosaic virus

ToMV: Resistance to tomato mosaic virus

TSWV: Resistance to tomato spotted wilt virus

V: Resistance to verticillium wilt

Those are some of the most common abbreviations for the most common diseases, but it's not an exhaustive list.

Planting the Vegetable Garden

We can garden year-round in all but the coldest (mountainous) regions in the Carolinas. We can plant almost everything directly into the garden, skipping the indoor seedling flats. I've even direct-sowed (planted right outside) tomato seeds in my zone 8a garden and had plenty of luck getting a large tomato harvest.

There are some plants that are easier to deal with outside as transplants, though. Whether you grow your own transplants or buy them, here's what you should plant outside as transplants and what you can grow from seed.

Seeds	Transplants
Bean	Basil
Carrot	Broccoli
Chives	Brussels sprouts
Cucumber	Cabbage
Dill	Cauliflower
Lettuce	Celery
Okra	Collards
Parsley	Eggplant
Parsnip	Kale
Onion	Leek
Pea	Pepper
Pumpkin	Spinach
Radish	Swiss chard
Turnip	Tomato
Watermelon	

Some plants are best grown from transplants, including the broccoli being planted by this gardener.

Starting Seeds Indoors

If you want to grow unusual varieties, you might have to order seeds and grow your own transplants. This isn't difficult, but it can be time-consuming. Look at the seed packets for information about when to start indoors. (You'll start seeds inside either a certain number of weeks before the last frost in spring or before the first frost in fall.)

Fill tray with seed-sprouting mix, and plant seeds according to package instructions.

Cover seeds until they sprout.

Supplies
Seed-starting mix
Seedling tray
Clear plastic cover
Heat mat
Grow lights

Purchase kits that have seedling flats and covers that fit, or use plastic wrap. Fill the seedling tray with seed-starting mix, not potting soil. This lightweight mix is sterile and specially formulated to encourage easy, problem-free germination. Plant the seeds according to package instructions, and gently water the seeds in the soil, taking care not to wash them away. You don't have to use a heat mat, but using one will result in faster germination and growth, especially for warm-season vegetables like tomatoes and peppers.

Keep the seeds covered until they sprout, then hang the grow lights 2 inches above the seedling tray. You don't have to use grow lights, but it is difficult to get strong seedlings without them. As the plants grow, move the lights up so that they are always 2 inches above the plants. Keep the seedling mix moist but not soggy. Damping off is a problem when seeds stay too wet and cold while germinating.

Planting Transplants Outdoors

Plant transplants outside according to the spacing the fully grown plants will need. Pay attention to the depth of the hole, and ensure that you don't bury the stem in the soil (except in special cases—see Growing Tip on the following page). Take the temperature of the soil to make sure it is warm enough. (The soil should be at least 60 to 65 degrees F for planting warm-weather vegetables such as tomatoes or peppers.) Before planting any transplants outside, prepare them by hardening them off.

GROWING TIP

Tomato plants should be planted deep. Strip off all but the top four sets of leaves. Plant the entire rest of the plant below the soil line. Tomato plants will grow roots from the stem, making them stronger and healthier.

Hardening off before Planting out

Vegetable transplants grown inside a greenhouse (or your house) need to be hardened off (acclimated to the change in temperature and light) before they're planted outside. Even if you buy plants that were sitting outside at a garden center, it's a good idea to harden them off before planting. For all you know, the plants were taken from the greenhouse, loaded on a truck, and brought to the garden center on the same day you saw them sitting outside.

How to Harden Off Transplants

1. Place plants in a sheltered location such as a porch or patio for the day, and bring them in at night. Do this for three or four days.
2. Next, leave them outside all day in the protected location. Do this for about a week. Don't forget to water while you're doing this!
3. Finally, move the plants from the sheltered location (the porch or patio) to a more exposed location (the front sidewalk or driveway). Leave them there for three or four days.
4. Wait for a cloudy day (if possible) and plant your plants in the garden. Planting out on a cloudy day will lower the stress that the plants experience.

Planting Seeds Outdoors

The key to success with seeds outdoors is to pay attention to them. Seeds have to stay moist while they are germinating. If they dry out, they will die—they have no reserves to draw on.

Harden off plants by setting them on your porch or patio during the day and bringing them in at night until they are acclimated to outside conditions.

Take the Temperature

Before planting seeds outside, measure the temperature of the soil to make sure it is warm enough. Radishes, for example, need soil temperatures of 50 degrees F or more.

Sow According to Finished Spacing and Germination Rates

Seed packets will have instructions to sow "thickly" (lettuce) or sow "thinly" (carrots, radishes). Usually these instructions correspond to germination rates. Radishes have a high germination rate, meaning that almost all of the seeds that you plant will sprout. Lettuce has a fairly high germination rate, but you want to grow lots of small leaf lettuce plants in a small space to encourage fast, leggy, tender growth. The more closely you can plant according to the finished spacing requirements, the fewer seeds you'll waste, and the less time you'll spend thinning.

Seedlings are fragile when they first sprout. Keep them moist, but not soaking wet.

Cover with Seedling Mix

Seedling mix isn't just for indoor plants. It's my secret weapon for growing seeds outdoors. Instead of covering seeds with garden soil, cover them with seedling mix. This mix will stay loose and resist soil crusting, making it easier for the seeds to sprout.

Check Seedlings and Water Frequently

Seeds need to stay moist but not soaking wet. Check them a couple of times a day. You might want to invest in a superfine hose nozzle to water without washing the seedlings away.

Further care for vegetables is discussed in Chapter 6.

Planting the Fruit Garden

Fruits are a more permanent part of the garden and landscape. Once you plant an apple tree, you're not going to move it. Because of their long tenure in the garden, it is also important to plant fruit trees and shrubs correctly to avoid long-term problems. Before digging any holes, check the soil pH. Certain fruits, such as blueberries, are picky about the pH of the soil (they need acidic soil, or soil with a lower pH).

Selecting Fruit Varieties

There are four main criteria to keep in mind when selecting fruit varieties:

1. Chilling hours needed to produce fruit
2. Height of mature tree or shrub
3. Disease-resistance properties
4. Pollination requirements

In the individual fruit profiles you will find recommendations for suitable varieties for Carolina gardens that favorably meet conditions for each criteria.

Planting Fruits

Before planting, test the soil pH, and adjust it if it is too high or too low.

Fruit plants can be purchased as container-grown or bare-root stock. Plant container-grown fruit trees and shrubs the same way you'd plant a landscape tree or shrub.

How to Plant a Container-Grown Fruit Tree or Shrub

1. Use a shovel or marking paint to mark the area for the hole. The planting hole should be at least twice as wide as the tree's rootball.
2. Dig the planting hole. This hole should be just as deep as the rootball— no deeper! If you sharpen the spade before digging, this step will go faster.

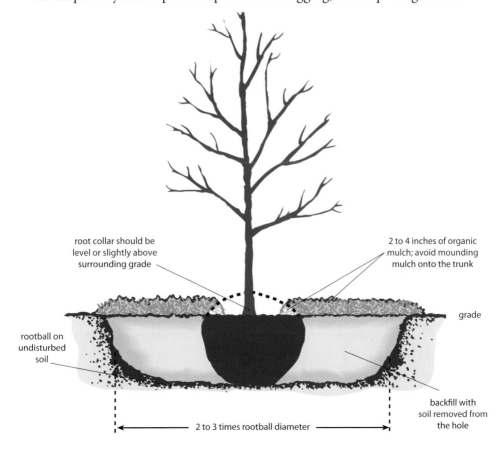

root collar should be level or slightly above surrounding grade

2 to 4 inches of organic mulch; avoid mounding mulch onto the trunk

grade

rootball on undisturbed soil

backfill with soil removed from the hole

2 to 3 times rootball diameter

GROWING TIP

Have you heard the saying, "plant 'em high"? Well, that refers to trees. Trees will settle a bit after planting. Always make sure that you finish the job with the top of the tree's rootball about 3 inches above the soil line. If you plant a tree too deep, the place where the tree trunk and the tree roots meet can rot, which will kill the tree.

3. Set the tree in the planting hole to check the depth. If the top of the rootball is lower than the soil line around the edge of the planting hole, add some soil back into the hole, pull the tree out of the pot, and replace the tree in the hole. You never want the crown of the tree (the part where the tree trunk meets the tree roots) to be below the soil line. In clay soils, set the rootball so it is a few inches above the soil line.
4. Fill in around the tree with the same soil that you removed from the planting hole. Do not add fertilizer or new topsoil. Water will move more easily and the tree will root properly if the soil in and around the planting hole is the same.
5. Mulch around the tree, taking care to pull the mulch away from the tree trunk. Do not create a mulch "volcano" around the tree (by piling mulch up high around the trunk)—that just encourages insects and creatures that snack on tree bark to take up residence next to your delicious young tree.
6. Water the tree. Plan to water newly planted trees every three days (every other day if it is hot and dry). New trees don't need to be staked unless they're in areas prone to heavy rains and frequent winds. It can take a couple of years for newly planted trees to root into the surrounding soil, so continue to monitor your tree for signs that it needs water.

How to Plant a Bare-Root Fruit Tree or Shrub
1. Soak the bare roots in water the night before you plan to plant.
2. Dig a hole twice as wide as the diameter of the root spread and about three inches deeper than the length of the roots.
3. Build a mound of soil in the center of the hole that is as high as the hole is deep.
4. Place the plant on top of the soil mound and spread the roots over the mound.
5. Fill in the soil around the plant, taking care that the point where the roots meet the trunk is above the soil line.
6. Finish by mulching around the plant.

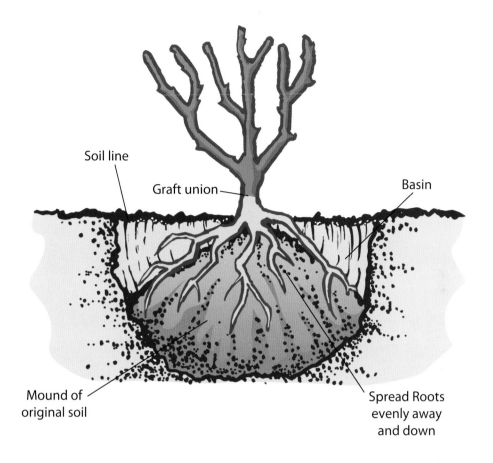

Soil line

Graft union

Basin

Mound of original soil

Spread Roots evenly away and down

The Rest of the Garden

Leave room in the vegetable garden for some flowers to attract pollinators. In addition to planting flowers, you can let some of your early-season vegetables bolt and flower. These flowers, blooming in spring and early summer, will attract pollinators to the first flowers on your vegetable plants.

The more diversity of plant life you have in the rest of your yard and garden, the healthier your edibles will be. Other plants will support beneficial insects and pollinators that contribute to the size and health of your harvest.

CHAPTER 6

GROWING YOUR GARDEN

Once the soil is prepped, the plants are purchased, and the garden is planted, the heaviest lifting is done, until it's time to replant to accommodate a shift in seasons. Throughout the growing seasons, you'll need to keep your plants producing prolifically by staying on top of these garden-maintenance tasks:

thinning	staking
pinching	fertilizing
watering	controlling pests
weeding	harvesting
mulching	succession planting

Mulching a vegetable garden with straw helps the soil retain moisture.

While the general principles of garden care apply to both fruits and vegetables, fruits have some unique issues and needs that are discussed in Part 2, preceding the individual plant profiles. You can take everything you learn from this section, add it to the additional information about fruit trees, and care for your fruit garden.

This garden was thinned when plants were smaller to allow for adequate space for the vegetables to grow and spread out.

Tools

CUTTING TOOLS

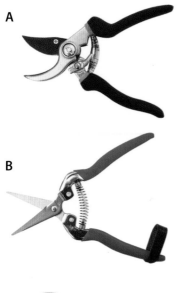

Hand pruners

A Hand pruners have a cutting blade and an anchor blade. They're good for pruning, harvesting vegetables, deadheading, and cutting bamboo stakes.

Snips

B Snips are a cross between scissors and hand pruners. They have a scissor-like cutting action, but they have a spring like hand pruners, for easy cutting. Snips are great for deadheading.

Scissors

C A good pair of scissors is a must for the garden. Use them to snip twine, deadhead, and harvest cut-and-come-again lettuce.

Velcro, twine, coated wire

D You always need something to tie a plant to a stake. Velcro and plastic-coated wire can be reused and work well with vegetables. Twine is invaluable because it is versatile. You can use it to create trellises for vining vegetables such as peas or cucumbers. Jute twine is the most commonly used type of garden twine, because it is strong.

PLANT STAKES AND SUPPORT

You can use anything long and narrow to stake plants. When you go to buy stakes, you'll run into a few common types. Here's more about each of the stakes and support pictured on page 51.

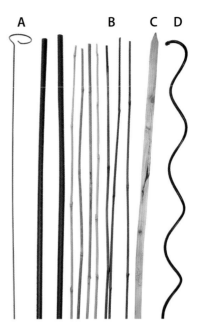

From left to right:

Metal or plastic-coated metal stake

A These heavy-duty stakes are perfect for vegetable gardening. They won't break as plants get tall and heavy.

Natural or dyed bamboo stake

B The light color of natural bamboo stakes makes them stand out more against dark green garden foliage, while the dyed-green stakes blend in. Bamboo stakes are easy to cut to size with hand pruners. They are best used for vegetables that stay fairly small. You can also use them to make a latticework for peas to climb up.

Wood stake

C Thicker wood stakes come in handy for staking peppers. Most people don't think of staking peppers, but they will grow taller and produce more fruit if they have some support.

Tomato spiral stake

D For use with tomatoes, which can twine around the spiral for support.

Tomato cage support

E Tomato cages are made of aluminum and provide support for tomatoes, peppers, eggplants, tomatillos, and any large vegetables.

DIGGING TOOLS

Pickahoe

A The scary-looking pickahoe actually makes planting easier and faster. This is a hoe with a sharp blade and a short handle. Use it as you would use a pickax to quickly plant vegetable transplants.

Trowel

B Trowels are small shovels used for weeding and planting. You can buy trowels made of plastic or made with metal blades and wood handles.

Soil knife

C Soil knives are indispensable. One side of the blade is sharpened, and the other is serrated like a saw. Use these to plant transplants, dig up weeds, and cut through roots. Some soil knives have depth measurements marked on them, which helps with planting. A soil knife is really a more useful version of a trowel.

Shovel (not pictured)

Shovels have long, straight wood handles and curved or straight bottoms. Use shovels to dig up plants, dig planting holes for trees and shrubs, and move compost or soil.

Hoe

D There are many different types of hoes. Hoes with narrow ends are excellent for digging planting rows. Hoes with larger, wider ends are helpful for spreading soil and mulching.

RAKING AND SPREADING TOOLS

Four-tine claw

A A four-tine claw has a long wood handle and a metal top with four curved tines. (This tool is sometimes called a cultivator.) Use a four-tine claw to cultivate and dig between vegetable garden rows, rake mulch in narrow garden beds, and incorporate compost and soil amendments into the garden.

Hard rake

B A hard rake is a cross between a leaf rake and a four-tine claw. It has hard metal prongs like a claw, but a wider head like a leaf rake. Use this to rake soil and spread mulch. Hard rakes are also useful for spreading gravel and raking mulch pathways.

Garden fork or pitchforks

C Pitchforks have long, straight handles and scooped heads similar to those of shovels, but they have sharpened tines instead of solid metal blades. Use pitchforks to turn the compost pile and move mulch around. Garden forks have handles like spades—with a grip you can put your hand through on the end—but they have flat tines that are wider thanpitchfork tines. Use garden forks to divide plants and turn over weeds.

WATERING TOOLS

Garden hose

A The garden hose is your main tool for watering plants. It's worth the money to buy a good 50-foot, heavy-duty, no-kink hose. You'll be glad you made the extra investment.

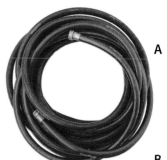

A

Soaker hose

B Soaker hoses are porous hoses that you connect to a faucet, place in the garden, and leave there. When you turn on the faucet, water slowly seeps out of the hose, reaching the plants where they need it most—their roots. Use soaker hoses if you don't have the time or money to create drip irrigation systems. Soaker hoses come in handy during short periods of hot, dry weather, because they're easy to move around. You can also get sprinkler timers to put on the ends of soaker hoses to create your own DIY irrigation system that will turn on and off while you're out of town. Tomatoes, especially, benefit from soaker hoses, because tomatoes crack or get blossom-end rot during periods of wet/dry/wet/dry watering.

B

Watering wand

C Every gardener needs a watering wand. This is a long tube with a water breaker on the end that disperses the water pressure from the hose. Watering wands make it easy for you to water near the base of a plant without bending over, and to reach across rows.

C

Water breaker

D Water breakers can be used with watering wands or on the ends of hoses. Sometimes called rosettes, water breakers disperse the flow of water so that it is gentler and less likely to cause soil to splatter. Water breakers should be used when you hand-water plants. They're especially good for watering newly planted plants and seedlings.

D

These carrot seedlings will need to be thinned to leave at least 3 inches between plants in order to harvest larger carrots later.

Garden Maintenance Activities

Calling these activities "tasks" or "chores" makes them sound a lot more difficult and time-consuming than they really are. The largest amount of work in vegetable gardening is getting the garden planted. The rest of it can be done in an hour or two after work or on the weekends.

Thinning

When you plant seeds, you usually plant more than you'll need if they all sprout. That leads to thinning, or removing some of the plants in the row to open up space for the plants you leave to grow. Save yourself work by sowing seeds with good germination rates fairly thinly and according to the recommended final spacing for the plants. If you don't thin root vegetables such as radishes, carrots, turnips, beets, and so forth, you'll end up with a lot of small roots instead of the fat, juicy roots you want.

Thinning is my least-favorite garden chore, and sometimes I don't do it. That has worked out fine for carrots, but not so well for radishes. To thin without disturbing the other plants, use scissors to snip the seedling off at the soil level.

Pinching

You can pinch plants such as tomatoes and peppers to get fewer, larger fruits. Pinching tomatoes also helps prevent these vigorous plants from putting energy into leaves rather than fruits. You can pinch the side shoots that grow directly between the tomato's main stem and the productive branches. You can also let these shoots grow from the main stem and pinch them after they have one set of leaves.

Pinching tomato suckers (the sprouts between the leaves and the stem) will lead to fewer, larger tomatoes being produced.

Watering

Watering is probably the most important technique to master in edible gardening. Vegetables and fruits only produce well if they are happy and not stressed. If they receive too much water, not enough water, or inconsistent water, they become stressed and can be afflicted by pests, diseases, and physiological problems. You can water vegetables by hand or by using irrigation.

Watering Seedlings

Seeds and seedlings need to stay continually moist while sprouting. They don't have reserves to fall back on. If you plant seeds in the garden, check them at least twice a day to make sure they are staying moist.

Watering by Hand

Hand-watering is a good thing to do if you want to relax at the end of the day or spend some time outside before going to work. Water by holding the hose nozzle at the base of each plant and counting to five.

GROWING TIP

Before you head into the house after work, or before you get in the car to go to work, do a lap around your garden. Water any plants that need to be watered, and keep an eye out for pests or anything amiss. If you set a time to do this, you won't let garden maintenance get ahead of you, and you'll give your plants the best care. (Plus, it's fun!)

Watering with Irrigation

Using irrigation or soaker hoses isn't a "fix it and forget it" option. You still have to monitor the plants to see if they need water and turn on the irrigation or set the timer. It's better to water all plants as deeply as possible and as infrequently as possible without stressing them. Don't let the irrigation run for five minutes, five times a day. Set it up to give the plants enough water to soak the top six to eight inches of soil. When that's dry, water again.

Watering wand

Lay a soaker hose after you plant, while the plants are still small. Use sod staples to hold the hose in place.

Weeding

Weeding is a constant battle, but by staying on top of it continuously, you can lessen the problems you have with weeds over time.

There are a few ways to deal with weeds:

- Use a pre-emergent herbicide to prevent weed seeds from sprouting. (This does not work on perennial weeds.)
- Hand-pull or dig weeds with large taproots (such as dandelions).
- Turn over the soil or rake under small weeds. (This works best for annual weeds.)
- Use herbicide to kill the weeds.

Why Bother to Remove Weeds?

Weeds are unsightly, for one thing, but they also take sunlight, nutrients, and water from the plants you want to grow. It's worth it to remove them.

Seasonal Weeds

In the Carolinas, we have two seasons of weeds to go along with our two seasons of edible gardening: winter and summer. You can use pre-emergent herbicides, which are herbicides that keep weeds from sprouting, to control annual weeds that spread by seed. Preen is a common synthetic pre-emergent. For these treatments to work, they have to be applied at least a month before the change of seasons. That means February to March for summer weeds and August to September for winter weeds. Read label

WHAT IS A WEED. AND WHAT IS A SEEDLING

Sometimes seedlings look like weeds. In this picture, you can see nasturtium seedlings (light green circular leaves) sprouting along the edges of the raised-bed box. In the center of the photo, next to the lettuce, are dollarweed plants (dark green circular leaves growing flat in the raised bed), which look like nasturtiums. Know what you're removing before you remove it.

Dollar weed masquerading as nasturtium seedlings.

directions carefully when using these products to determine how long you must wait before you can seed new crops.

WEED TYPES

The following weed types are not "official" distinctions but rather categories of weeds you'll have to deal with. Here are the characteristics of certain types of weeds and how you deal with them.

Annual

Annual weeds complete their life cycle during one season by growing, flowering, and setting seeds. The main objective in preventing annual weeds from spreading is to prevent them from flowering. You can mow or cut these weeds frequently, pull them out, or use herbicide to kill them. Pre-emergent herbicide works well on annual weeds.

Perennial and Biennial Clumping

Perennial clumping weeds come back every year. Biennial weeds grow a large plant with a taproot one year and then flower and set seeds the next year. Both types will eventually flower and set seeds, which establishes new clumps. These types of weeds respond well to hand-digging, because they do not have runners or stems that can break off and start new plants. Just be sure to get the whole root since most of these weeds can regrow from root pieces left in the soil.

Perennial Spreading

Perennial spreading weeds are the most difficult to control. Dollarweed is the chief offender, but there are others. These weeds are most easily eradicated with systemic herbicides, which are weedkillers that are absorbed by the plant and taken to the roots to kill the plant from the inside.

Mulching

Mulching the vegetable garden will save you a lot of time watering and weeding. It is amazing the difference that a thick, 3-inch layer of mulch will make. If you mow your own lawn and don't use chemicals on it, save the grass clippings—they make great mulch, and they add nutrients to the soil. Just remember to let them age for a few weeks before spreading and make sure they don't have seed heads.

Other good materials for vegetable garden mulch include:

Wood mulch in a vegetable garden.

Shredded newspaper
Shredded bark mulch
Aged manure
Compost
Wheat straw
Shredded leaves

There's a misconception that you shouldn't use wood mulch in a vegetable garden. Now, you wouldn't want to use treated wood mulch or sawdust, but shredded hardwood mulch is more beneficial than detrimental. If you can buy shredded hardwood mulch with compost in it, even better!

When mulching around your plants, avoid mounding the mulch up around the stems of the plants, which can cause the plants to rot.

Staking

You'll stake certain vegetables to let them grow upright and reach the light. You'll stake other vegetables because their brittle stems can't support the weight of their fruits.

Vegetables That Need Trellises
Certain vegetables will grow up trellises (wood, metal, or string). Certain plants need to have trellises to grow.

Vegetables That Can Grow up Trellises
Cucumber
Pumpkin
Squash

Vegetables That Must Grow up Trellises or Lattices
Pole bean
Garden pea

Vegetables That Need Stakes
Eggplant
Okra
Pepper

Vegetables That Need Cages
Tomato

A word from experience: It is much easier to place the plant supports before (trellises) or right after (everything else) planting than it is to try to wrestle a large, viny plant into a cage or support after it has started growing.

Fertilizing

Vegetables and fruits need a constant supply of nutrients in order to produce food for us to eat. Different plants have different requirements, which are listed in the individual plant profiles. In the Carolinas, adding compost or worm castings to the soil at the time of planting will help provide nutrients for the plants, but our natural soils are so fast draining and nutrient deficient to start with that you can't get away without adding fertilizer, too.

I always stake my peas by making latticeworks with sticks that fall out of the trees in the yard over the winter.

When tying a plant to a stake, line the stake up to the main stem of the plant. Then tie the plant to the stake, but never tie the string so tight that the plant is smashed against the stake.

Sidedress plants with organic fertilizers (sprinkle them on the soil alongside the plants) every few weeks to keep your vegetables growing.

Fertilizers all have a combination of nutrients in them. Most of them have the "Big Three" macronutrients: nitrogen (N), which promotes green leafy growth, phosphorus (P), which strengthens roots and flowers, and potassium (K), to help with flavor and hardiness. Some fertilizers have micronutrients such as calcium, magnesium, manganese, and iron. Fertilizer labels include what is called an "analysis," which is the percentage of each of the "Big Three" nutrients contained in the product, listed in order of N-P-K.

DIFFERENT TYPES OF FERTILIZER

You might have grown up with Miracle-Gro, a common brand-name synthetic fertilizer, but you have other choices.

Organic or Natural Liquid Fertilizer

Organic or natural liquid fertilizers are usually made from kelp, fish emulsion, or a combination of each. Mix this type of fertilizer with water in watering cans to water into the soil. Some organic or natural liquid fertilizers are used as foliar feeds. Mix these fertilizers with water in a pressurized sprayer, and spray the leaves of the plants. The plants will soak up the nutrients through their leaves. Some organic fertilizers are smelly, so don't use them right before you have an outdoor barbecue. Organic fertilizers tend to have more micronutrients along with the "Big Three." These fertilizers will build your soil health over time.

Conventional or Synthetic Liquid Fertilizer

Synthetic liquid fertilizers have nitrogen, phosphorus, and potassium in them in one combination or another. These fertilizers are formulated for different types of plants. You can buy houseplant fertilizers, fertilizers for acid-loving plants, vegetable fertilizers, and fertilizers formulated to encourage more blooms on plants; these have higher levels of P. You can buy conventional fertilizer in concentrated liquid or powder forms. In both cases, you need to mix the concentrate with water. Some brands of fertilizer have special attachments that will do the mixing for you. Synthetic fertilizers usually just have macronutrients: nitrogen, phosphorus, and potassium. They will not build soil health over time.

Slow-Release Fertilizer

Slow-release fertilizer feeds plants over a period of three to five months. The nutrients in the fertilizer are pelletized in a form that breaks down and releases nutrients to plants over time. Sometimes a slow-release fertilizer is combined with pre-emergent weed control in a "weed and feed" product. Don't use weed and feed products where you've just planted seeds, or they won't sprout.

Slow-release fertilizers include time-release synthetic products (for example, Osmocote) and organic fertilizers (for example, Plant-tone). Their nutrients are typically released over two to four months. These products do not have to be reapplied often—usually only at planting time and again two to three months later.

Nutrient Deficiencies

If your plant leaves are turning weird colors (purple, yellow), they might have nutrient deficiencies. Nutrient deficiencies cause highly predictable results, and it's usually possible to diagnose whether a nutrient deficiency is a problem by looking at the plant.

The diagram below shows the symptoms caused by the most common nutrient deficiencies.

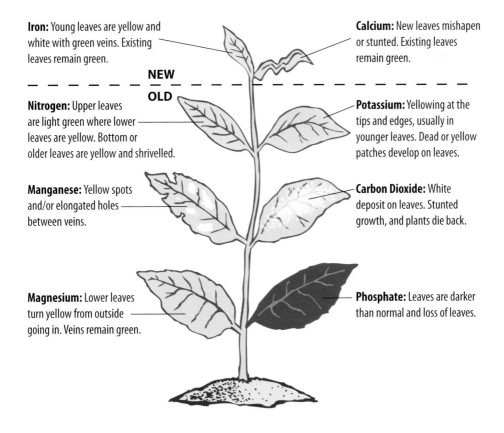

Iron: Young leaves are yellow and white with green veins. Existing leaves remain green.

Calcium: New leaves mishapen or stunted. Existing leaves remain green.

NEW

OLD

Nitrogen: Upper leaves are light green where lower leaves are yellow. Bottom or older leaves are yellow and shrivelled.

Potassium: Yellowing at the tips and edges, usually in younger leaves. Dead or yellow patches develop on leaves.

Manganese: Yellow spots and/or elongated holes between veins.

Carbon Dioxide: White deposit on leaves. Stunted growth, and plants die back.

Magnesium: Lower leaves turn yellow from outside going in. Veins remain green.

Phosphate: Leaves are darker than normal and loss of leaves.

GROWING TIP

If you see nutrient deficiency problems in your plants, test the soil pH. If the pH is too high or too low, it won't matter if you fertilize—the plants won't be able to get the nutrients from the soil.

Controlling Pests

There is nothing more discouraging than carefully tending your garden only to see it being chewed apart by pests. Each type of edible has its own pest problems, which are outlined in the individual plant profiles. There is also a chart of pests and diseases on page 65 to help you identify what the problem is and how to fix it. Before you take action, it's important to understand pests, pesticides, and their relationship to your garden.

Pests: A Fact of Life

There are insects and bacteria and fungi everywhere, eating everything, including each other. On a single tree in the forest, there could be four hundred types of caterpillars, four hundred types of insects eating the caterpillars, and thousands of bacterial organisms eating the insects. What this means is that ecosystems have their own checks and balances. When we use chemicals to control certain pests, we are messing with the checks and balances, and that can have consequences.

Diversity and Pest Problems

I hardly ever have to use pesticides in my garden, primarily because I have a high tolerance for pest activity, and I also grow a lot of different types of fruits, vegetables, flowers, trees, and shrubs in my yard. When you grow a biologically diverse garden and have a biologically diverse landscape, you support the beneficial insects that prey on harmful insects.

The way you lay out the vegetable garden matters in terms of pest control, too. If you plant all of your eggplants in one spot, and all of your tomatoes in another spot, and all of your beans in another spot, you are giving pests an all-you-can-eat buffet they can enjoy without having to leave their chairs. Try planting clumps of each edible throughout the garden, intermixed with flowers. The pests might find one clump and stick with it, allowing the other plants to thrive.

Beneficial Insects

You can encourage beneficial insects, such as ladybugs, lacewings, soldier beetles, wasps, and others, to live in your yard by limiting the use of broad-spectrum (kills everything) insecticides and by growing a diverse landscape.

Remember, insecticides that kill everything kill the insects that are working for you, too.

Integrated Pest Management

If you're serious about growing all of your own food and you can find the time, I highly recommend taking pesticide training at your local Cooperative Extension. Not only will you learn how to safely apply pesticides, but you'll learn about integrated pest management, or IPM. There are four pillars of IPM:

Set action thresholds: You don't have to spray at the first sign of an insect. Decide what amount of damage you're willing to live with.

Monitor and identify pests: Early detection leads to early action before much damage can be done.

Prevention: Do what you can to prevent pest and disease problems. Plant resistant varieties, use barriers to keep pests from reaching the plants, and provide plants with the best care possible so that they can withstand pest problems.

Control: After all of the other steps, you can decide how to control the pests. Controls can be organic or synthetic, involve treating the pests or pulling out the affected plants.

Organic versus Synthetic Controls

There's a misconception that if you use an organic pesticide, you're not putting yourself or your plants in any danger. That is false. There are organic pesticides made from naturally occurring ingredients that are just as deadly as pesticides that are made in a lab. When deciding to use a pesticide to control weeds, pests, or diseases, always read the label and follow the instructions for dosage, protective equipment, reentry timing, and pre-harvest windows.

Controlling Mammalian Pests

Deer, rabbits, chipmunks, and squirrels are all problems in the edible garden. The most effective way to control these pests is with barriers. Fencing, repellents, and netting are just about the only way to keep these pests out of the garden. If you're growing vegetables in rows, think about using row covers to keep pests away.

For more information about individual pests, see the chart on page 65.

Harvesting

You've prepared the soil, planted the plants, fertilized them, tended them, and kept other creatures from eating them. It's time to harvest! But *when* do you harvest?

It's easier to tell when some fruits and vegetables are ready than others. Information about harvesting each plant type is found in the individual plant profiles. Here are some general tips about harvesting:

- Let fruits and vegetables that you want to be sweet ripen on the vine or plant for as long as possible. If you pick a tomato green, it will turn red and soften, but it won't be as tasty.
- Use clippers or scissors to harvest eggplants, peppers, and other fruits that don't easily snap off. You don't want to yank the plant out of the ground.
- Harvest in the morning—that's when everything has the highest water content, which means it will last longer before eating.
- Learn how big the finished product is supposed to be, so that you know when to harvest. If the eggplant is supposed to be 1 inch wide and 4 inches long, and you let it grow to be 4 inches wide and 1 foot long, it will not taste good.

Some vegetables are easier to harvest with scissors, so you don't yank the plant out of the ground. Here, the gardener is using scissors to cut the pepper. If she twisted the pepper off the plant, she could break or damage the rest of the plant.

Troubleshooting

Plant	Pest Problem	Symptoms	Solution
Apple	Apple scab	Hard scabby growth on apple fruits and on leaves	Plant resistant varieties, use fungicides according to label
	Cedar apple rust	Yellow spots with red rings on leaves	Remove junipers and crabapples, plant resistant varieties, prune off galls from junipers and cedars
	Fire blight	Wilting and dried up tips and leaves	Cut off infected parts and throw away, plant resistant varieties
	Codling moth	Holes in fruits	Pheremone traps, hygiene (pick up fallen fruits), insecticides
Arugula	Flea beetles	Presence of insects	Use row covers directly after seeds have sprouted or seedlings have been planted
Basil	Snails and slugs	Holes in leaves	Diatomaceous earth
Bay	No pest problems		
Bean	Thrips and aphids	Presence of insects, dry spots on leaves	Insecticidal soap
	Bean beetle, corn earworms	Presence of insects	Row covers after planting, insecticides labeled for specific pests
Beets	Downy mildew, leaf spot	Grey-white fuzz	Fungicides (treatment not usually necessary)
Blackberry	Birds	Fruits half-eaten	Netting
	Crown borers	Wilting of new growth	Cut off affected area below the wilting point
Blueberry	Birds	Disappearing fruit	Netting
	Cranberry fruitworm	Nets in the shrubs	Cut off affected area and throw away
Broccoli	Cabbage looper	Moths flying around at night, holes in the leaves, presence of caterpillars	*Bacillus thuringiensis (B.t.)*
Brussels Sprouts	Aphids	Presence of insects, sticky honeydew produced by ants farming the aphids	Insecticidal soap
	Cabbage loopers, cabbage worm	Presence of insects, holes in the leaves	*Bacillus thuringiensis (B.t.)*
Cabbage	Cabbage loopers, cabbage worm	Presence of insects, holes in the leaves	*Bacillus thuringiensis (B.t.)*
	Black rot	Black tips on leaves, brown patches	Plant resistant varieties
	Flea beetles	Presence of insects	Row covers after planting
Carrot	Parsley worm (black swallowtail caterpillars)	Presence of insects	Control not usually necessary
Cauliflower	Slugs and snails	Presence of slugs, holes in the leaves	Diatomaceous earth
	Cabbage root maggots	Wilting plant	Row covers immediately after transplanting
Chives	No pest problems		
Cilantro/ Coriander	Aphids	Presence of insects, sticky honeydew on plants	Blast with water, use insecticidal soap
	Whiteflies	Presence of tiny white insects	Sticky traps
Citrus	Scale	Presence of small, black insects (look like ticks), mildew around scale	Keep the plant outside in stress-free conditions
Collards	Aphids	Presence of insects, sticky honeydew on plants	Blast with water, use insecticidal soap
	Cabbage worm, cabbage looper	Presence of insects, holes in the leaves	Row covers after transplanting

(continued on page 66)

Troubleshooting (continued)

Plant	Pest Problem	Symptoms	Solution
Corn	Birds	Tops of ears ripped apart	Netting
	Armyworm, corn earworm	Holes on the plants	*Bacillus thuringiensis (B.t.)*
Cucumber	Cucumber beetles	Presence of insects	Floating row covers
	Squash vine borer	Sudden wilting of plants	Floating row covers and Sevin for prevention (nothing you can do once they strike)
Dill	Tomato hornworms	Presence of worms, entire plant defoliated	Handpick and destroy
Eggplant	Tomato hornworms	Presence of worms, entire plant defoliated	Handpick and destroy
	Colorado potato beetle	Orange eggs on underside of leaves, insects present	Crush eggs, use row covers after planting
Fennel	Black swallowtail caterpillars	Caterpillars present	Control not necessary
Fig	Root knot nematode	Test soil before planting	Plant resistant varieties
	A variety of blights	Off-color leaves	Plant resist varieties, prune to allow air movement within tree
Garlic	No pest problems		
Grape	Fungal diseases	Spots on leaves, mildew on fruit	Prune to promote good air circulation
Hardy Kiwi	Root knot nematodes	Test soil before planting	Fumigate soil prior to planting if nematodes are present
Kale	Aphids	Presence of insects	Blast with water, insecticidal soap
	Cabbage worms	Holes in leaves, presence of insects	*Bacillus thuringiensis (B.t.)*
Kohlrabi	Flea beetles, cabbage worms	Holes in leaves	Floating row covers
Leeks	No pest problems		
Lettuce	Aphids	Presence of insects	Blast with water
Loquat	Fungal diseases	Black spots on leaves and fruit	Fungicides
Marjoram	Aphids and spider mites	Presence of insects	Keep plant stress free--water in times of drought
Melon	Cucumber beetles	Presence of insects	Floating row covers
	Squash vine borer	Sudden wilting of plants	Floating row covers (nothing you can do once they strike)
Mint	No pest problems		
Okra	No pest problems		
Onions	Onion root maggots	Wilting tops	Diatomaceous earth
Parsley	Black swallowtail caterpillars	Presence of insects, chewed-on plants	Do not control. They will go away.
Parsnip	Black swallowtail caterpillars	Presence of insects, chewed-on plants	Do not control. They will go away.
Peach and Nectarine	Plum curculio	Sticky sap coming out of fruits	Insecticides applied at petal fall of flowers
	Numerous other diseases	Vairous	Consult local cooperative extension for a spray plan
Pear	Fireblight	Ends of branches look burned	Plant resistant varieties
	Pear psylla	Presence of insects, brown spots on leaves	Dormant oil during winter

Troubleshooting (continued)

Plant	Pest Problem	Symptoms	Solution
Pea	Aphids	Presence of insects	Spray with water
Pecan			Pest control not practical, plant varieties recommended for southeast
Pepper	Viruses	Off-colored leaves, wilting	Plant resistant varieties
Persimmon	Few pest problems		
Plum	Plum curculio	Sticky sap coming out of fruits	Insecticides applied at petal fall of flowers
	Brown rot	Yellow growths on plum, shriveled appearance	Fungicides
	Black knot	Warty growths on branches	Cut off affected and dispose of branches
Radish	No pest problems		
Rosemary	Mealybugs and scale	Whitish downy fungus on plant, black spots on plant	Horticultural oil
Sage	Slugs	Presence of slugs, holes in leaves	Stop watering the plant
	Spider mites	Red dots on leaves	Blast with water, give the plant extra water
Spinach	Leaf miners	Brown trails and dried sections of leaves	Floating row covers after planting
	Fungal problems	Rotten leaves	Keep water off leaves
Strawberry	Botrytis	Moldy fruits, flowers	Fungicides
	Birds	Holes in fruits	Nets
Summer Squash	Cucumber beetles	Presence of insects	Floating row covers
	Squash vine borer	Sudden wilting of plants	Floating row covers (nothing you can do once they strike)
Sweet Potato	Sweet potato weevil	Tunnels under the skin	Plant resistant varieties
Swiss Chard	Aphids, leaf miners	Presence of insects, dried spots on leaves	Blast with water, insecticidal soap
Tarragon	No pest problems		
Thyme	Aphids and spider mites	Presence of insects	Blast with water, give extra water
	Fungal problems	Rotten leaves and stems	Stop watering
Tomatillo	Cutworms	Dead seedlings	Wrap bottom three inches of stem in newspapwer
	Tomato hornworms	Presence of worms, defoliated plants	Handpick and destroy
Tomato	Tomato hornworms	Presence of worms, defoliated plants	Handpick and destroy
	Various viruses	Discolored leaves, wilting plants	Start by planting resistant varieties
Turnip	Root maggots	Wilting tops	Row covers to keep adults from laying eggs
Winter Squash	Cucumber beetles	Presence of insects	Floating row covers
	Squash vine borer	Sudden wilting of plants	Floating row covers (nothing you can do once they strike)

PART II

FRUIT & VEGETABLE PROFILES

By now, you've clued into the fact that this book is organized a bit differently than many books about edible gardening. It isn't a straight alphabetical list. It isn't organized entirely by type. To make it easier to grow these edibles, the book is organized around the way these plants grow—where they grow in the garden and which season you plant and harvest.

First is a discussion of the fruits, organized by plant type: fruit and nut trees, citrus, shrubs and brambles, and vining and "garden" fruits. The size and type of fruits determines where you plant them, when you harvest them, and which varieties will work best for your garden.

Blueberries forming on 3-year-old blueberry shrubs.

Botanical Names and Common Names

All plants (all living things) have a scientific name composed of a genus (first word) and species (second word). Those names are standard throughout the world. The scientific names of plants are also referred to as botanical names, and the language of them is known as botanical Latin.

Solanum lycopersicum is the botanical name for tomato. The botanical name is always italicized. Cultivars of tomato include 'Brandywine' and 'Cherokee Purple', among others. Cultivars are always unitalicized and in single quotes.

Because different cultivars or varieties grow well under different circumstances, it's important to do your research before shopping. I've included recommended varieties or cultivars in the individual plant profiles.

In a few cases, mostly with fruit plants, there are several species of a fruit within one genus. Some species grow better in the Carolinas than others. Blueberries are one example of this. That is also noted in individual plant descriptions.

Next up are the vegetables and herbs, organized by cool and warm season. In the Carolinas, the most influential factor governing vegetable and herb growing is the season. Certain plants will grow only during the cool season, and others during the warm season. Still others confound us by tolerating both seasons equally well.

While fruit trees and shrubs are a permanent fixture in the landscape and thus have to be selected and planted accordingly, most vegetables and herbs are grown as annuals and can and should be moved around. Fruit tree care centers on growth for longevity, while vegetable care is oriented toward getting the biggest harvest for one season.

At the beginning of each section, there's information pertinent to planting just those types of (or seasonal) plants. Ideally you can take the book to the garden center or home-improvement store, get what you need to grow a garden for the season at hand, go home, plant, and start growing.

Planting flowers around your vegetable garden will add color and attract pollinators.

Find Your Favorites

You can check the index at the back of the book, but you can also use the alphabetical list below to find your favorite edibles.

CHAPTER 7

FRUIT & NUT GARDENING

Planting a fruit tree or shrub is like making an investment in the garden. It can take as many as ten years to get a harvest from a nut tree. Most fruit trees bear (produce) fruit somewhere between two and six years after planting. When you buy fruits, the source (website, catalog, garden center) should be able to tell you when the tree will bear. By the time you buy some plants (citrus, particularly), they're already of bearing age and you won't have to wait.

Peach trees laden with fruit.

Choosing Fruits and Nuts

First thing's first: nut trees are fruit trees. Nuts are the fruits of the trees that bear them, just like a zucchini squash is the fruit of a zucchini plant, and a tomato is the fruit of a tomato plant. From this point, as there is just one nut tree in this book, all of the fruit trees and nut trees will just be referred to as "fruit trees." Their care is similar enough that it won't make a difference.

Fruit trees and shrubs (and vines, for that matter) take up a lot of space in the garden relative to vegetable and herb plants, and they're more permanent. It's important to choose plants that produce fruits that you like to eat,and to be mindful of space. Most fruit trees need full sun in order to produce well. If you're starting with a blank slate of a landscape, the sky can be the limit. If you are trying to work fruits into an existing landscape, siting is trickier.

There are many factors that influence which fruit plants you choose for your garden. When selecting fruits to grow at home, you'll need to pay attention to these characteristics:

Source
Tree size
Rootstock
Bloom time
Pollination requirements
Pest and disease resistance
Chilling hours
What you like to eat
Whether you'll eat everything fresh or preserve part of your harvest

One is not more important than the other. All of the above characteristics influence how a fruit plant grows and whether you'll be successful growing it.

Source

Thoughtfully selecting the source of fruit plants is essential to successful fruit growing. You'll have more long-term success if you purchase locally grown fruit plants than if you purchase plants grown in greenhouses or fields on the other side of the country. Different weather conditions, different pests, and different diseases affect fruit trees much more than plants such as tomatoes or green beans that live for one season and then are replanted.

Pecans, blueberries, apples, and peaches are all types of fruit plants that should be acquired from local sources due to various problems with pest and disease resistance. Be sure to select varieties that are adapted to your location (climate and pests). If you purchase online, these varieties should be labeled for growing specifically in the Southeast. Just because a plant is hardy in a particular growing zone doesn't mean that it will be well suited to growing in your area. There's much more to think about with fruits than cold tolerance.

Plant Size

The largest fruit plants that you can grow are nut trees. Apples follow close behind, if they are not grafted on dwarfing rootstock. Pears, plums, and peaches are the next smallest (in that order). Vining fruits such as muscadine grapes and hardy kiwis are about equal with shrubby fruits such as blackberries and blueberries in terms of space requirements. Melons take up a lot of space but can be easily moved from year to year. Before buying a particular fruit, decide whether you truly have the space for its eventual size and if you want to commit that space over a period of ten to twenty years.

The swelling on a tree trunk indicates the graft union where the rootstock (bottom) and scion (top) of a fruit plant grow together.

Rootstock

The ultimate size of many fruit plants depends on the rootstock. Most fruit trees have two parts. The top of the plant, which produces the fruit that we eat, is called the scion. Rather than growing from seed, scions are cuttings from other plants of the same variety. (This process was the original "cloning.") The bottom of the plant is called the rootstock. Rootstocks can cause a scion to grow more slowly, increase cold hardiness, shorten the time between planting and fruit production, and add resistance to pests. The scion is grafted onto the rootstock, and the two pieces of plant grow together to form one plant. The point where these two join is called the graft union.

In the Carolinas, rootstocks are more valuable for their ability to add cold hardiness or disease resistance than for their ability to control size (which is discussed in the "Growing Tree Fruits and Nuts in the Carolinas" section on page 83). Root-knot nematodes are a problem in the coastal areas of the Carolinas. Growing peaches, nectarines, and plums on resistant rootstocks is the only way you'll have success with those plants. (Specifics for plant selection are listed with each plant profile.)

A fruit tree orchard shows trees in various stages of blooming.

Bloom Time and Pollination Requirements

Fruit plants have to be able to flower in order to produce fruits. Most fruit plant flowers need to be pollinated, and some need to be cross-pollinated with pollen from another variety.

When selecting fruit trees and shrubs for your home orchard, pay attention to whether a plant is self-fertile or self-pollinating (in which case you can plant just one plant and still get fruit), or if the plant requires cross-pollination (in which case, you need at least two plants). If you have limited space, go for self-fertile plants.

For trees requiring cross-pollination, bloom time also matters. If one variety of tree or shrub is finished blooming before the one you bought to cross-pollinate it, neither will bear fruit. Apples, peaches, pears, and other flowering fruits are categorized into groups based on blooming times. You need to plant at least two from each group in order to get good pollination and fruit set.

Pay attention to pollination requirements when purchasing fruit plants. Otherwise, you could find yourself with healthy plants that never produce a harvest.

Pest and Disease Resistance

Fruit trees and shrubs are susceptible to an alarming number of pest and disease problems. There are sprays and treatments that you can use to get rid of the problems, but you can give yourself a head start by choosing resistant plants. In grafted fruit plants, both the scion and the rootstock play a part in disease resistance. Pay attention to the recommendations in individual plant profiles.

Chilling Hours

Fruit growers north of the Mason-Dixon Line have an easier time getting good fruit harvests, as long as the plants are cold hardy where they live, than fruit growers in the South. Blame it on chilling hours—or hours when the air temperature is below 45 degrees F. Temperatures above 60 degrees F reverse chilling hours accumulated. Most fruit trees have chilling hour requirements that correspond to the length of time a plant will remain dormant before flowering. Chilling hours are required in many plants to trigger a break in dormancy, leading to blooming.

If you plant a tree that requires 500 chilling hours in an area that receives only 300 hours, the plant will not flower and will not produce fruit.

Chilling hours are discussed in more detail in individual plant profiles. Recommended plant varieties for each area of the Carolinas (coastal, Piedmont, or mountains) have been selected, in part, with chilling hours in mind.

End Use

Fruit trees and plants need a lot more care, for a much longer time, than other edibles. When picking out what you're going to plant, think about what you like to eat. Many gardeners find themselves branching out and trying new things if those new things come from their own gardens, but you want to be realistic. How many figs can you eat? There's no reason to plant six plants unless you're going into business making fig jam for everyone in your city. You don't need twelve apple trees if you're not going to put up quarts and quarts of apple butter.

Only plant things you like to eat, and only plant trees that will bear a manageable harvest for you.

GROWING TIP

If you have more fruit than you know what to do with, check with local food pantries, homeless shelters, and meal programs. Some cities have "gleaning" programs, where volunteers will come and pick your extra harvest and take it to a needy charity prepared to make use of it. Other food banks can accept and distribute your fresh produce. Don't let fresh food go to waste.

Apples ready for harvest.

Planning a Fruit Garden

How many of each fruit do you plant? How much room do you need? The number of plants depends on the size of the plant, the yield you hope to harvest, and whether the plants require cross-pollination.

You can start small with two apple trees, four to six blackberry plants, four to six blueberry plants, and twelve strawberries.

Or you can tuck a few fruit trees along your lot line, add a row of blackberries near the vegetable garden, and devote a couple of rows or a raised bed from the vegetable garden to strawberries.

For me, the more fruit I can grow, the better. I have *never* had a problem giving away extra homegrown anything. The mistake most fruit gardeners make is not planting enough of each type of plant, thus limiting (or downright preventing) good pollination and fruit set.

My only exception to this rule is figs, which are prolific producers. Nobody needs more than one fig plant. They're like rabbits, only they're plants.

Caring for Fruits

Each type of fruit plant has its own specific schedule of care, but there are maintenance tasks that are important to all fruit plants. You will need to make time to take care of these tasks in your fruit garden.

Parts of a Fruit Tree

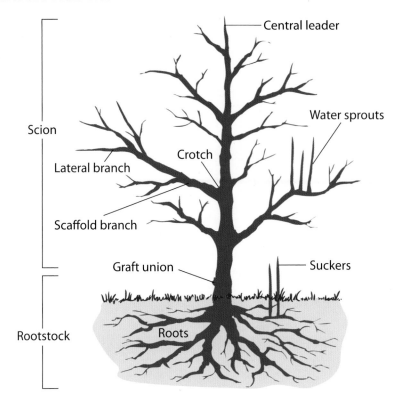

To understand the directions in the plant profiles, you need to be able to identify the parts of a fruit tree.

Rootstock: The bottom of the tree; the roots.

Scion: The top of the tree; the portion that produces fruit.

Graft union: The place on the tree where the rootstock meets the scion.

Central leader: The main trunk of the tree. (Apples and pears are pruned with a central leader. Peaches and plums are not.)

Water sprout: A branch that grows straight up from a scaffold branch on the fruit tree. Water sprouts do not produce fruit and should be removed.

Scaffold branch: A main horizontal branch from which fruiting spurs or branches grow.

Lateral branch: A smaller branch growing off of a scaffold branch. Lateral branches grow out, not up, and produce fruiting stems called spurs.

Crotch: The angle of a scaffold branch to the tree. Wider crotch angles are stronger and produce less breakage than narrow angles.

Suckers: Sprouts coming up from the rootstock. Sometimes suckers come from the stem below the graft union, and sometimes they appear to sprout right out of the soil. They should always be pruned off. Never put systemic weedkiller on a sucker. You can kill the whole tree.

Pruning

Pruning isn't just for fruit trees. Muscadine grapes, blackberry bushes, and blueberries also require pruning. You'll prune to control the size of the plants, to get rid of old growth and encourage new growth, to increase sunlight penetration into the tree canopy, and to allow for airflow. Pruning helps establish the architecture of fruit trees and the renewal of blackberry patches.

Most pruning is done during winter when plants are dormant. A few plants benefit from summer pruning as noted in the individual plant descriptions.

PRUNING TOOLS

Here are the tools that you need in order to prune:

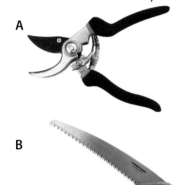

Hand pruners

A Hand pruners have a cutting blade and an anchor blade. They're good for pruning, deadheading, and cutting bamboo stakes.

Pruning saw

B Pruning saws fold up for easy toting in the garden basket. They are handy for cutting branches over 2 inches in diameter.

Loppers

C Loppers are like big hand pruners. They have the same type of cutting blade, but usually it is larger. Loppers also have longer handles. They're good for trimming shrubs and tree branches that are out of arm's reach.

Pole pruning saw

D If you grow semidwarf or standard-sized fruit trees, a pole pruner with both a lopping and a saw attachment is a handy tool. You'll use the saw just like you'd use a pruning saw, only with a longer handle. The lopping attachment will have a string to pull to engage the cutting blade. With a pole pruner, you can reach higher up into the tree without having to climb a ladder.

PRUNING TECHNIQUES

Once the plants are in the ground, the most crucial aspect of fruit tree care is pruning. Individual instructions for fruit tree pruning are listed in the plant profiles. Here are the basics.

Why Prune?

- To encourage new growth
- To let light and air into the tree
- To establish good structure (which makes it easier to harvest fruits)
- To control size

Fruit trees are usually pruned to have a central leader with scaffolds (apples, pears) or in an open-vase shape (peaches, plums). The scaffold method establishes a central leader with branches that decrease in length as you move up the tree. The open-vase shape establishes several main lateral branches that bear fruit. You can use heading cuts, thinning cuts, and renewal pruning to establish these shapes.

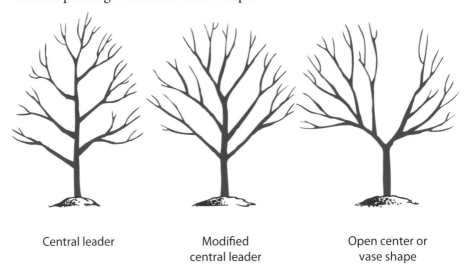

Central leader

Modified
central leader

Open center or
vase shape

Fruit trees are shaped by pruning. Apples should be pruned in a pyramid-like shape, while peaches and plums are pruned in an open, vase shape.

TYPES OF PRUNING

Heading cuts: When you cut off the end of a branch, that's called a heading cut. The response of the plant will be to produce more side-shoots below where you made the heading cut.

Thinning cuts: When you remove some of the "bulk" or interior branches by cutting them all the way back to the main stem, you're making thinning cuts.

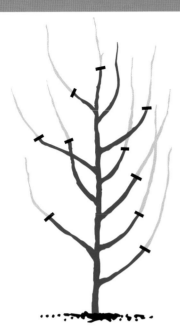

Heading cuts involve cutting off the ends of plant branches.

Thinning cuts involve removing some of the interior growth of the plant.

Renewal pruning: This is a process most often used with shrubs, but which can also be applied to fruit trees, whereby you remove at least one-third of the old growth on the plant each year, stimulating new growth.

Pruning stimulates fresh new growth that will produce fruit. Old orchards are brought back to life by careful pruning of the old trees. Remember: When you prune fruit trees, it's like cutting hair—the plants will grow back.

Renewal pruning is the process of removing one-third of the plant's growth each year.

Thinning: Fruit trees require thinning in order to produce larger fruits. The trees will self-thin about halfway through summer. You'll know this has happened, because suddenly a bunch of little fruits will be lying all over the ground. You can go through after this and continue to selectively thin. Pick fruits so that there is only one fruit every 4 to 6 inches along a branch.

How to Remove Large Branches

When you need to remove a large branch (anything that requires a pruning saw), use the following three-cut pruning technique.

1 Make the first cut on the underside of the branch, about 6 inches from the tree trunk. You'll only cut one-fourth to one-third of the way through the branch.

2 Make the second cut farther out on the branch from the first cut. Cut the branch all the way off. The branch will probably break off while you're cutting it. That is why you made the first cut on the underside of the tree—to help the branch break in the direction you want without stripping the bark off the branch.

3 You can see the finished second cut, here. At the very bottom of the cut edge, you can see where part of the branch ripped. Do not make a cut that strips the bark off of a tree branch.

4 Cut off the branch stub remaining on the tree. Place the pruning saw just outside of the branch collar, which is the bark swelling between the branch and the main trunk. Saw through to remove the stub. Do not cut the branch flush with the tree trunk, or you'll hurt the tree's chances of healing itself.

5 In this picture, you can see that there's still about ¼ to ½ inch of branch left to allow the branch collar to heal. Never cover pruning cuts with tar, concrete, or sealant. The tree will heal itself if left alone. Sealant or tar creates a dark, moist environment that is perfect for bacteria to grow in. Sealing a cut can actually hurt the tree. If you follow this pruning technique, it will heal itself.

Staking and Support

Most landscape trees do not need to be staked when first planted, but many fruit trees do. Blackberries, muscadine grapes, and hardy kiwis require some sort of trellis to climb up for support. Fruits can be shallow rooted, which makes them easily dislodged by wind. Plan to stake newly planted trees and to construct sturdy trellising systems for newly planted vining fruits.

Other care considerations unique to different types of fruit plants are discussed in more detail in the individual plant profiles.

Pest Control and Harvest Protection

Fruit pest control depends upon the pest and the host plant. Some pests are controlled with a spray of horticultural oil when the plants are not leafed out. Other pests have to be controlled while fruit is developing on the trees.

Pest control on fruits never stops, because observation of the trees and plants never stops. Different pests are active at different times of the year. Make it a habit to look at your fruit tree orchard. If something seems amiss, take a cutting or take some pictures and identify the problem. If you're having trouble, consult your local Cooperative Extension office.

If birds are a problem, use netting to keep them from carrying away your harvest before you can.

GROWING TIP

Fruit trees are highly temperature sensitive.

Extremely cold temperatures can freeze flower buds. Some plants can tolerate more cold than their flower buds can, which means the trees might live through a cold snap, but the flowers won't. No flowers equals no fruit.

Temperature fluctuations also cause problems—mostly for plants that need to be cross-pollinated with pollen from another plant. The Carolinas are notorious for temperature fluctuations. It can be 70 degrees F one day and 25 degrees F two nights later. These temperature fluctuations can cause plants to break dormancy and flower too early. If you have three apple trees and one tree flowers early while the other two trees flower normally, the early-flowering plant can miss being pollinated. If all three plants pollinate during a freakishly warm stretch of early March when none of the insects that pollinate them are active, they will miss being pollinated and the flowers may freeze.

Dwarf pear trees are easily managed from ground-level, as they grow no taller than 10 feet or so.

Growing Tree Fruits and Nuts in the Carolinas

Growing fruit trees is as rewarding as it can be frustrating. To get good fruit tree growth and robust harvests requires patience (it can take a while for plants to bear fruit) and vigilance (there are always insects and diseases waiting in the wings to destroy your harvest).

Choosing the right variety is an important part of success in fruit tree growing. Another factor in selecting a tree is deciding upon the size that you want.

You can purchase standard trees grown on seedling rootstocks. These fruit trees can grow from 20 to 30 feet tall. Most home gardeners look for semidwarf and dwarf fruit trees. Semidwarf trees (10 to 20 feet at maturity) are best, if you have the space. They are usually longer lived, establish deeper roots, and have more disease resistance than dwarf trees (5 to 10 feet at maturity).

Aside from the natural height of the tree (apples are naturally taller than peaches), the rootstock usually determines the size of the tree.

Find Your Favorite Fruits and Nuts

You can check the index at the back of the book, but you can also use the alphabetical list below to find your favorite edibles.

Standard pear trees grow much larger.

GROWING TIP: PLANTING NEW FRUIT TREES

When planting new dwarf or semi-dwarf fruit trees it helps if you can stake the trees. These trees are grafted onto rootstocks and can be more fragile than non-grafted "seedling" standard-sized trees. Staking the trees will keep them from blowing over in the wind as they're establishing their root systems. It helps to stake these trees for up to two or three years (re-setting the wires yearly to allow the tree to grow without wires cutting into the bark). What happens if you don't stake? They will blow over. I speak from experience. All of my nectarine trees are now growing at an angle because I was too lazy to stake them or to pull them up after a hurricane! But, it doesn't take a hurricane to knock them over. If you want healthy trees, start them right by staking.

APPLE *(Malus domestica)*

With the right selection of variety, apples can be grown in the Carolinas.

You'll never want to eat a store-bought apple again after you taste one that you harvested from your own tree. They're sweeter, crunchier, juicier, and have so much more flavor! Your own apples also won't have the waxy coating that commercial growers put on their fruits to keep them "fresh" during transit.

Even though the Carolinas are warm and most areas have relatively few chilling hours (temperatures below 45 degrees F) in comparison to other, more northern states, we can still grow apples. The key is in selecting varieties that are resistant to fireblight and require the appropriate amount of chilling for your region in order to produce fruit.

All apples that we grow today are actually two-part plants. The top of the plant, which produces the fruit we eat, is called the scion. Rather than growing from seed, scions are cuttings from other plants of the same variety. The bottom of the plant is called the rootstock. Rootstocks can cause a scion to grow more slowly, increase cold hardiness, shorten the time between planting and fruit production, and add resistance to pests.

When you purchase an apple tree, you select not only the scion (or top), but often a particular rootstock as well. Nursery catalogs sometimes offer different choices—allowing you to purchase, say, a Gala apple with a dwarfing or semidwarfing rootstock. A "standard" tree is one that grows to full size without the benefit of a dwarfing rootstock. The size that you select depends on how much space you have to grow the tree, the type of soil in which you will be planting it, and the average low temperature for your area.

▥ Recommended Varieties

When selecting apple trees, you need to think about three things:
- ▥ The preferred size (and thus the preferred rootstock)
- ▥ Bloom time for pollination
- ▥ Disease resistance, particularly fireblight

Rootstocks Suitable for the Carolinas

Semidwarf rootstocks will produce trees that are about two-thirds the size of a standard tree. Common semidwarf rootstocks for the Carolinas include MM.106, MM.111, and M.7. These rootstocks cut the time from planting to fruit production from ten years down to four to six years.

Dwarf rootstocks produce trees that are half to one-third the size of a standard tree and reduce time from planting to fruiting from ten years to two to four years. Common dwarf rootstocks for the Carolinas include M.9 and M.26.

Rootstocks are helpful for growing apple trees in the Piedmont and the mountains. Only standard seedling trees should be grown in coastal areas.

Planting Trees for Cross-Pollinating

In order to get apples, you need to plant at least two trees that bloom at the same time so that they can cross-pollinate each other. Some trees are sterile and do not produce fertile pollen. If you plant one of those trees, you need to plant at least two others in order to get apples on all of the trees.

This chart shows apple varieties suited for our area, and trees that work well as pollinators. Trees marked with an asterisk (*) have sterile pollen.

Apple Tree Varieties and Cross-Pollinator

Variety characteristics in order of maturity

Variety	Fruit Color	Fruit Use	Relative Bloom Time	Potential Cross Pollinators
Gala	Yellow-orange to red	Fresh	Early to midseason	Golden Delicious
Empire	Dark red over green background	Fresh, cooking	Early	Golden or Red Delicious, Gala
Jonagold*	Yellow with light red stripes	Fresh, cooking	Midseason	Gala, Empire
Golden Delicious	Yellow-green to light yellow	Fresh, cooking	Midseason to late	Red Delicious, Gala
Red Delicious	Red	Fresh	Early	Golden Delicious, Gala
Stayman*	Blush to red	Fresh, cooking	Midseason	Gala, Golden or Red Delicious
Rome	Blush to red	Fresh, cooking	Late	Fuji, Braeburn
Braeburn	Green with light red blush	Fresh	Midseason	Rome, Fuji
Fuji	Green with red stripes	Fresh	Midseason	Rome, Braeburn
Anna	Red blush	Fresh	Early	Dorsett Golden
Dorsett Golden	Yellow	Fresh	Early	Anna
Arkansas Black	Dark red	Fresh, cider	Midseason	Braeburn

*Tree has sterile pollen.

Varieties for Different Areas of the Carolinas

Coastal regions: 'Anna' and 'Dorsett Golden' each need only 300 chilling hours in order to bear fruit. (That's amazingly low.) These varieties are well suited for the warmest areas of the Carolinas. Varieties that can be grown in northern coastal areas include 'Red Delicious', 'Golden Delicious', and 'Granny Smith'.

Piedmont: 'Red Delicious', 'Golden Delicious', 'Gala', 'Jonagold', 'Granny Smith', 'Ginger Gold', 'Fuji', 'Stayman', 'Arkansas Black', 'Braeburn'

Mountains: 'Rome', 'Arkansas Black', 'Gala', 'Red Delicious', 'Gala', 'Jonagold', 'Fuji', 'Granny Smith', 'Braeburn'

Planting

Apples are tree fruits and are planted the same way you would plant any tree. Refer to the instructions on page 45. You can plant apple trees at any point during the year, although it is best to avoid planting them during the worst heat of summer (mid-June through August).

Also important in terms of planting is spacing. Apples need to be planted close enough to each other for pollination. Plant standard trees no more than 75 feet apart, semidwarf trees no more than 45 feet apart, and dwarf trees no more than 25 feet apart. Plant trees so that they have room to grow without touching at their mature spread. (That varies from tree to tree. Read the label for expected dimensions.)

When planting apple trees, make sure that the graft union (where the scion meets the rootstock) is at least 5 inches above the soil. Stake grafted trees for the first two to three years to ensure that they take root and, literally, stay together while getting established in the garden.

Maintenance

Once planted and staked, the primary maintenance concern is pruning. You need to water the trees as they are becoming established and fertilize yearly in spring by spreading a balanced slow-release fertilizer according to package instructions around the root zone of the tree.

To allow trees to establish themselves well, remove all fruit that forms during the first three years. (Difficult, I know, but your trees will be healthier in the long run.)

Pruning is the critical type of maintenance for apple trees, though. From the moment you plant a tree, the way you prune it and train it will determine whether it is healthy and has good structure for bearing fruits or whether it is susceptible to breakage in high winds, breakage from heavy fruit loads, or low fruit production from lack of light.

Pruning to Establish New Trees

This diagram shows how to prune a tree purchased from a nursery:

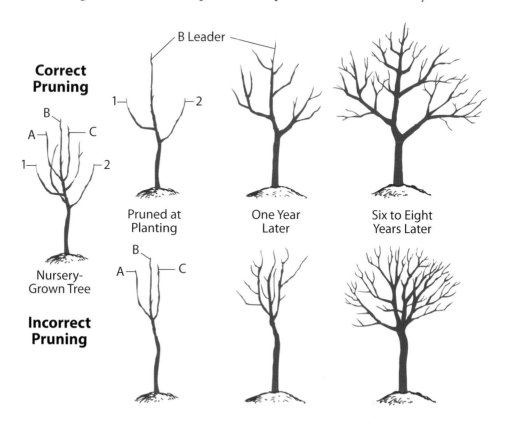

When pruning fruit trees, the goal is to establish good architecture. This allows air circulation in the canopy (top of the tree), permits light penetration (sunlight must reach the inner branches for fruit production), and creates a branch structure that can withstand wind and the weight of fruit on a branch.

Pruning to Maintain Established Trees

Once the tree has become established (four or more years), you'll need to prune to maintain good architecture and encourage fruiting. Always remove water sprouts, which are branches that grow straight up from side branches. These aren't growing in the right direction to produce good long-term shape.

Apple tree spur

In addition to structural maintenance, you'll prune to encourage fruiting. Some apple trees produce fruits from spurs, which are small, compressed, nubby stems along the branches. It's a rule of thumb that the fewer fruits there are on the tree, the bigger the fruits will be. If there are too many spurs on the tree, you can use hand pruners to simply snip off the spurs at even intervals. If the tree doesn't have enough spurs, you can cut off the end of the branch, which will cause the plants to sprout more spurs along the branches. Some apples bear fruit from the tips of branches. Prune these trees by cutting back the ends of branches by 8 to 10 inches, leaving five or six buds on the branches.

The overall guiding principle of pruning all fruit trees is to maintain a somewhat triangular shape. This allows light to reach every part of the tree. (Remember that no light equals no fruit.)

■ Pest Problems

Apples are members of the rose family (along with many other edible fruit plants) and are susceptible to a wide variety of pests and diseases. Variety

Cedar apple rust

selection is key in growing healthy apple plants, as is identifying plants in your garden that could be harboring pests and diseases. Crabapples, junipers, and arborvitae all harbor the cedar-apple rust pest. If you want to grow apples in your garden, you might consider getting rid of these three ornamentals.

Other apple problems include apple scab, powdery mildew, and fireblight. Try to plant cultivars that are resistant to these diseases. Spraying dormant oil during winter can control some insects, including aphids and mites. Fruit worms, including the Oriental fruit moth and codling moth are the worst insect pests for apples. Insecticide sprays are needed to control these pests. The two most important times to spray are as soon as the flowers fall from the petals and again two weeks later. Never spray insecticides on open blossoms—this will kill pollinators.

Apple maggots are another pest. To control them, hang sticky traps—available in garden centers and home-improvement stores—in the trees. You can also protect individual fruits by placing a plastic bag around each

Apple scab

fruit and attaching it with a clothespin. (NOTE: Horticultural oil mainly controls insect pests, not diseases. It can control powdery mildew, but must be sprayed in the growing season when the disease is active to do this.)

If you have trouble identifying what pest, disease, or cultural problem is plaguing your apple tree, take pictures, including close-ups, and cut a branch and take it to your county extension service.

Harvesting

Apples are ready for harvest when they easily come off the tree in your hand. Don't rip apples from the tree, as you can damage the spurs or buds that will produce fruit next year.

GROWING TIPS

Apple trees will self-thin in early summer. Do not be alarmed when you see some of the fruits drop off the tree. It's a natural process that will result in larger fruits ripening in fall.

Pick up dropped fruits, and rake and remove leaves each year.

Certain apple diseases will overwinter in dead leaves and fruits.

Control pests by practicing good garden hygiene.

LOQUAT *(Eriobotrya japonica)*

You'll usually see loquats growing as ornamental plants in the warmer (non-mountainous regions) of the Carolinas, but they will bear fruit in zones 8–10, which include most of South Carolina and the coastal regions of North Carolina. As an added bonus, loquats are moderately deer resistant, which means that's one less pest you have to worry about attacking your fruits. Loquat trees grow to a height of 10 to 15 feet in the warmest areas, and they bloom and produce fruit during winter.

■ *Recommended Varieties*

'Tanaka' is the most cold-hardy variety of loquat. Other varieties that reliably produce fruit include 'Gold Nugget', 'Champagne', 'Oliver', and 'Wolf'. In order to ensure good fruiting, plant at least two plants for cross-pollination.

■ *Planting*

Plant loquats in full sun to partial shade in well-drained soil. The trees can be slow to root, so stake them for the first year after planting. Loquats are one of the few trees that will produce fruits in partial shade, so if your yard is sun challenged, go for loquats.

Loquats will bear fruit in the warmer areas of the Carolinas.

GROWING TIPS

Loquats will tolerate a hard pruning. If these trees outgrow their space, just cut them back after they are finished fruiting.

Loquats are evergreen, but they lose leaves all year. To keep pests and diseases from attacking the plants, clean up leaves immediately after they fall off the tree.

Maintenance

Loquats are low-maintenance plants. If your area is experiencing a drought, give the plants at least 1 inch of water per week. In rich soils, do not fertilize, as extra fertilizer can lead to weak growth. Trees growing in sandy soils will benefit from an application of slow-release fertilizer in spring.

Pest Problems

Loquats are relatively pest-free, but they can be affected by fireblight, a bacterial disease that affects all members of the rose family.

Harvesting

Loquats bloom in the fall. Fruits are ready to harvest about 90 days after flowering in early spring. Ripe fruits are sweet, but unripe fruits are highly tart and acidic. The color of the ripe fruits differs from cultivar to cultivar, so check to see when the fruits on your tree should be ripe and what the ideal color will be. Use a knife or hand pruners to cut individual fruits from the tree.

Eat only the peeled, seeded fruits. The seeds and skin of loquats contain a chemical called amygdalin, which when eaten in somewhat small amounts, can cause lethargy and poisoning.

Loquat fruits ripen in the late winter or spring.

PEACH & NECTARINE *(Prunus persica)*

Homegrown peaches and nectarines are so delicious, sweet, and juicy because they can be left on the tree until they are ripe, instead of being picked hard and unripe to withstand shipping. However, both peaches and nectarines are more difficult to grow than most other fruit trees. They are pickier about pruning and pest control and are susceptible to a huge variety of pests, diseases, and cultural problems. Eating a peach that you harvest yourself is worth the effort, though. Just be prepared, if you plant peaches and nectarines, to put in some extra work to get the reward.

▆ *Recommended Varieties*

California-grown peaches are not suitable for Carolina gardens. When selecting your plants, it is imperative that you source varieties that have been specifically bred to grow well in the Carolinas.

Peach

Rootstocks

Rootstock selection is as important as scion selection. These rootstocks do well in the Carolinas:

'Lovell' and 'Halford': These rootstocks are time-tested in the Carolinas but are not nematode resistant.
'Nemaguard': Best for trees grown in coastal areas, as it is resistant to root-knot nematode problems.
'Guardian': A newer rootstock that is resistant to ring and root-knot nematodes. It is also good for coastal areas.

Varieties

Peaches: 'Redhaven', 'Norman', 'Winblo', 'Contender', 'China Peach' (white fleshed), 'Cresthaven', 'Encore', 'Legend', 'Flordaking'
Nectarines: 'Crimson Gold', 'Sunglo', 'Redgold', 'Durbin', 'Fantasia', 'Karla Rose', 'Armking'

Whenever possible, consult with local Cooperative Extension agents to see which varieties they recommend. You can buy a peach or nectarine tree anywhere and enjoy the flowers, but you can't buy one anywhere and expect to get fruits.

▆ *Planting*

Peaches and nectarines are self-fertile, so spacing is not as critical when planting these trees. Peaches grafted on dwarfing rootstocks are somewhat of a novelty. They cause plants to stay fairly small—most dwarf peach and nectarines won't grow any wider than 8 feet, but also bear few fruit. Plant along the edge of your property or as specimen plants.

Peaches and nectarines need well-drained soil to establish healthy root systems. They are not at all tolerant of heavy, poorly drained soils.

Follow instructions for planting trees on page 45, and stake peaches and nectarines for the first two years after planting.

■ Maintenance

Peaches are fast growers that do not need extra fertilizer, except in sandy soils. The biggest aspects of peach maintenance are pruning, thinning, and pest control. Prune peach trees to maintain an upright vase shape. Peaches produce lots of side branches, about half of which need to be removed yearly in order to channel energy into production of fewer, larger fruits. Peaches will self-thin in summer, dropping as many as half of the fruits. You can further thin peaches, leaving one peach or nectarine to mature every 4 to 6 inches along the branch.

Nectarine

To keep peaches productive (they produce fruit on one-year-old branches), prune after the plants have finished fruiting and you have finished harvesting to remove any water sprouts.

■ Pest Problems

Give yourself a leg up on pest control by planting resistant varieties. Even this won't spare your plants from all pests. If you want to grow fresh peaches and nectarines, you will have to follow a pest-control program. Whether you use organic or synthetic pesticides is up to you, but you won't have peaches without being vigilant about pests.

Diseases are a bigger problem than insects with peaches and nectarines. Peach leaf curl, brown rot, peach scab, bacterial spot, and other diseases are problems with peaches. Apply fungicides during the growing season to control these diseases.

■ Harvesting

Peaches are ready for harvest starting in early summer and continuing through late summer, depending on the variety. The fruits should easily come off the tree without yanking, and they should be slightly soft to the touch.

PEAR *(Pyrus* spp.)

We can grow both European and Asian pears in the Carolinas. Asian pears have rounder, firmer fruits than the European varieties. Both types are prolific producers—a couple of trees is enough for the home gardener. It can take up to nine years for pear trees to bear fruit, though varieties grown on dwarf rootstocks will bear fruits earlier. Asian pears will often bear two to three years after planting—another reason to try these unusual fruits.

Espaliered pear tree

Recommended Varieties

European: 'Moonglow', 'Orient', 'Magness', 'Seckel', 'Keiffer'
Asian: 'Shinseiki', 'Hosui', 'Kosui', 'Nititaka', 'Chojuro'

Planting

Plant pears in full sun in well-drained soil. Plant two or more trees of different varieties, 15 to 20 feet apart, so that they can cross-pollinate each other. While some pears are partially self-fertile, they'll produce a bigger harvest if they have a cross-pollinating tree nearby. Do your research before planting one of each, as opposed to two of one and two of another.

Maintenance

Pears, once established, do not need extra water. They rarely need extra fertilizer except in extremely poor sandy soil. As with other fruit trees, pruning is the most important aspect of maintenance (beyond pest control).

Prune pears in late winter in a similar manner to apple trees—to encourage airflow and light penetration. Maintain a triangular shape with a strong central leader (single central trunk) and scaffold (layers of branches that progressively get shorter toward the top of the tree).

Pears bear fruit on three- to ten-year-old spurs. Do not prune to remove spurs, as making more pruning cuts will open more areas for diseases to attack the plants. You will want to renewal prune by removing at least

10 percent of the canopy each year to promote new growth. Also be sure to remove water spouts and root suckers.

Pears bear fruit in clusters. To encourage production of fewer, larger fruits, thin fruits when they're 1 inch in diameter, leaving one fruit per cluster with clusters spaced every 6 inches along each branch.

Pest Problems

Fireblight is a big enough problem with pears that it is important to plant resistant varieties. You'll know that your tree has a fireblight problem if the leaves on specific branches turn brown and dry up starting at the tip, seemingly overnight. It's difficult to control this disease. You can control insect problems that affect pears, such as scale and mites, by spraying trees with dormant oil during winter.

Harvesting

Pears will be ready to harvest 100 to 140 days after flowering, depending on the variety. Allow Asian pears to ripen fully on the tree and then pick them. Asian pears will keep up to three months in the refrigerator. Harvest European pears when they are still firm to the touch and allow them to soften indoors.

It can take several years for a pear tree to bear fruit.

GROWING TIPS

Pear trees can have narrow branch angles that make the branches susceptible to breaking from high winds or heavy fruit loads. You can train the branches to have a wider angle by hanging weights on the individual branches when they are young and flexible. That sounds weird, but it works.

If you have space along a wall, you can espalier a pear tree to grow flat against a trellis.

Keep in mind, some pears can take 10 to 12 years to start bearing fruit. Be patient.

PECAN *(Carya illinoinensis)*

Pecans are in the hickory nut family and thus have a fond place in my memory. One of the houses I lived in as a child came with a fairly mature yard and garden. There were two big hickory trees in the front yard. I always looked forward to playing "cook in the kitchen," making mud pies with the nuts. Occasionally I tried to crack them and eat them, too—quite a task for a seven-year-old. To have success growing pecans in the Carolinas, make sure to plant varieties that are resistant to diseases found here. You also need to plant at least two trees of different cultivars to get the best nut harvest. Pecans thrive in the coastal and Piedmont areas of the state but not in the mountains.

▪ Recommended Varieties

Plant one plant from the type I group and one plant from the type II group for best nut yield.

Type I: 'Cape Fear', 'Pawnee'
Type II: 'Stuart', 'Sumner'

Container-grown trees are easier to establish than bare-root trees and can be planted at any time. Do not purchase a container-grown tree that is larger than 5 or 6 feet tall. Pecans establish taproots as they grow, which makes them hard to transplant. Bare-root trees should be planted outside from late fall to early winter.

Pecan trees are large, full-sized landscape trees.

▪ Planting

Pecan trees are large, so plant them where they have room to grow to heights of 40 to 50 feet. Be careful to avoid planting under power lines. Plant pecan trees 60 to 80 feet apart.

■ Maintenance

Mulch trees with compost, and water frequently (at least 10 to 15 gallons per week) when trees are first establishing themselves. Test the soil where pecan trees are growing to ensure proper levels of zinc and soil pH. Zinc deficiency can cause bronzing and mottling of leaves; early defoliation; dead twigs in tops of trees; abnormally small nuts; small yellowish leaves; and short, thin twigs growing on older scaffold branches, with rosettes of small, yellowish green leaves at the tips.

■ Pest Problems

Squirrels are one of the worst pests affecting pecan trees. This is primarily because they eat your harvest before you can. To protect trees from squirrels, wrap the trunk with a metal shield, 5 feet off the ground. You can attach the shield with spikes that you withdraw slightly each year as the tree grows.

Pecan scab is a major disease problem for pecans. You can limit the spread of this disease by cleaning up leaves, twigs, and nuts from the previous year and throwing them away (not composting).

Pecan weevil is the most serious pest of pecans. These grubs eat the nut and then exit through a small hole drilled in the shell. Pecan weevil is difficult to control. Contact your local Cooperative Extension office for recommendations of insecticides and application times.

Pecans can be eaten fresh or used in cooking and baking.

■ Harvesting

Start harvesting nuts when they start to fall. Lay a clean sheet under the trees and shake the branches to loosen the nuts. Once you harvest them from the tree, lay them in a single layer in a warm, dry place to dry out. You can freeze nuts in resealable plastic bags until you're ready to use them.

'Great Wall' persimmon is an Asian type.

PERSIMMON *(Diospyros virginiana* [American], *Diospyros kaki* [Asian]*)*

There are two types of persimmons: American and Asian (or Oriental). A huge, native American persimmon tree grew over the sidewalk next to the tiny house where I lived in college, across from the Horticulture Building, at Purdue University. I wish I had taken advantage of the fruits by cooking them instead of cursing the mess they made when we tracked them into the house. American persimmons have to be picked when ripe for best flavor. Asian persimmons can be picked when slightly underripe.

▓ *Recommended Varieties*

American persimmons require a male and female tree planted near each other in order to produce fruit. Asian persimmons are self-pollinating but produce heavier crops when cross pollinated.

GROWING TIP

It's a myth that persimmons have to withstand a frost before they can be harvested. In fact, frost will damage fruits.

Cold-hardy Asian persimmons include 'Great Wall', 'Korean', and 'Sheng'. Other recommended varieties include 'Fuyu', 'Jiro', and 'Hanagosho'. Native American varieties that grow well in the Carolinas include 'Early Golden', 'John Rick', 'Killen', 'Miller', and 'Garretson'.

Planting

Persimmons can be messy. Select a site where the tree will not grow over a driveway or walkway—unless you want to track the mess into your house. Plant cross-pollinating native varieties at least 50 feet apart so that the two trees don't crowd each other. Plant Asian varieties 15 to 20 feet apart.

Maintenance

Persimmons are drought tolerant and low maintenance once established. Asian varieties are smaller and require pruning similar to apple tree pruning.

Pest Problems

There are few pests that affect persimmon trees, but, by and large, it isn't feasible to control problems on a 40-foot tree. If the trees are grown in a diverse landscape, most problems will naturally resolve themselves.

Harvesting

Pick fruits in fall when they're soft. Nonastringent Asian varieties can be harvested when they are still firm.

One persimmon tree will bear a lot of fruit.

PLUM *(Prunus spp.)*

Buy a plum at the grocery store and it is likely to be hard and flavorless. Grow and harvest your own and you'll change your tune about eating these fruits. There are two main types of plums: European and Japanese. Dried plums, or prunes, are European plums. Japanese plums are juicy with thin skins. Excellent for fresh eating, these are the kinds most readily available in the grocery store. Asian plums are better suited to our warmer climate, though European plums thrive in mountainous regions of the Carolinas.

■ Recommended Varieties

Plums require a cross-pollinator tree of the same type. Japanese plums cross-pollinate other Japanese plums, and European plums cross-pollinate European plums. The 'Methley' variety of Japanese plums is self-fruitful and doesn't require another variety for pollination.

Japanese: 'Methley' (partially self-fruitful), 'Byrongold', 'Morris', 'Ozark Premier'
European: 'Bluefre', 'Stanley', 'Shropshire'

Before settling on a variety, consult with your local Cooperative Extension office. Plums can be grafted onto rootstocks that are susceptible to nematodes—a problem in the Carolinas. If you are growing plums in coastal areas, look for plants grafted onto 'Nemaguard' or 'Guardian' rootstocks. For growing plums in the Piedmont or the mountains, look for plants grafted onto 'Lovell', 'Halford', or 'Guardian' rootstocks, which are resistant to bacterial canker infections.

■ Planting

Plant plums in full sun, 18 to 22 feet apart in soils with a pH of 6.0 to 6.5. Plums are tolerant of different soil types, growing equally well in heavier but well-drained clay soils and sandy soils.

■ Maintenance

Plums benefit from being fertilized once a year in spring right after they finish blooming. Because they're heavy fruiters, these trees benefit from regular water during summer while fruit are swelling.

Prune plums in much the same way that you'd prune peaches—to let light into the trees, encourage air circulation, and create good branch scaffolding. Prune plums in late spring (after February 1).

Fresh, home-grown plums make excellent eating.

▓ *Pest Problems*

Plums are susceptible to brown rot, a fungal disease. Use fungicides labeled for use on peaches to treat this disease during the growing season. Black knot is a disease that causes warty growths on branches. Cut off affected branches and throw away the pieces. (Do not compost diseased plant parts.) Plum curculio is the worst insect pest affecting plums. There are insecticides that will control this pest. Always follow instructions when using pesticides.

▓ *Harvesting*

Plums are ready to harvest when the fruits have a white, waxy coating. Allow European plums to ripen fully before harvesting. Asian plums can be picked when they are slightly underripe. Be careful not to yank the fruits off the trees, because you don't want to break the fruiting spurs, which will bear again the following year.

GROWING TIP

The biggest difference in growing plums from growing other fruit trees is that you need to prune them to favor longer, outward-growing branches as opposed to a strictly triangular shape. This results in an open center, allowing plenty of light penetration.

GROWING CITRUS IN THE CAROLINAS

Potted lemon tree

Believe it or not, you can grow your own citrus in the Carolinas. The varieties suited to growing outside (due to cold hardiness) increase in number the farther south you go in the region, but in all parts of the Carolinas, you can easily grow citrus trees in pots.

Citrus plants fall somewhere between the fruit tree and fruit shrub categories. You can grow them in the landscape away from the house as small trees if you live in zone 9 or higher. Otherwise, you're more likely to grow them next to the house as shrubs or in containers that you can move around.

In the cooler areas (and even the northern coastal areas), you might find yourself throwing a blanket over the plants or bringing them inside for a few nights here and there because of a frost in the forecast, but that's doable. (Until you have more than ten or so plants. My husband loves growing citrus, so I speak from experience here. We fight for room in the garage—his plants compete for room with my surfboards.)

■ *Citrus Cold Hardiness*

Whether you can grow a citrus plant outside in the ground or in a pot that can be brought inside depends on where you are and what your average minimum temperature is. Citrus trees grown in pots are less cold hardy than their counterparts grown in the ground, so keep that in mind. If a plant tag says a tree is hardy to zone 9, if you're growing it in a pot, it will likely only withstand cold temperatures associated with zone 8.

◼ Sources

As with other fruit trees, if you can purchase plants grown in or near your area, you're more likely to have luck. Those plants are adapted to the growing conditions in your yard. Of course, if a plant is greenhouse grown, the location is moot. There are some nurseries that specialize in growing citrus in the Carolinas. Talking to the growers can also help you learn tips and tricks about growing citrus in your specific area.

◼ Rootstocks

Citrus varieties are usually grafted onto a rootstock. The most commonly used rootstock is the trifoliate orange. If you see water sprouts or suckers from the rootstock (which will grow from the bottom of the tree below the graft union), cut them off. Trifoliate oranges are not tasty to eat!

Trifoliate orange tree—source of rootstocks.

◼ Recommended Varieties

The most cold-hardy citrus that you can grow outside in the Carolinas (down to zone 8b) is the satsuma tangerine (*Citrus reticulata*). There are different named varieties of this citrus, but it is generally hardy down to 15 degrees F, though trees benefit from protection when temperatures dip below 25 degrees. 'Ten Degree Tangerine' is a named satsuma variety that lives up to its name of surviving down to 10 degrees F. It tastes more like a sweet lemon than a tangerine. 'Owari' is another cold-hardy satsuma variety that will survive temperatures as low as 15 degrees F.

Meyer lemons (*Citrus meyeri*) can be pruned to stay small and are hardy down to 20 degrees F, so they are a good choice for lemons. They are also very prolific producers. We've harvested as many as forty lemons off of a 3-foot-tall tree grown in a container.

Kumquats are also fairly cold hardy and will survive to temperatures as low as 15 to 20 degrees F. 'Meiwa' is a sweeter variety of kumquat for the home garden.

While you'll end up bringing some of these plants indoors anyway, the more cold hardy they are, the less often you'll have to cover them or bring them inside during winter.

◾ *Planting*

Citrus trees grow best in full sun to partial shade. You have two choices when planting citrus: in ground or in pots. Here's what you need to know about both techniques.

Planting Citrus in Pots

You should plant citrus trees in pots that are twice as wide as their rootballs. This will allow the trees to grow undisturbed for up to four years, without needing to be repotted and moved up to a larger container. Plant citrus in well-drained potting mix. You might consider mixing potting mix with some perlite for better drainage.

Planting Citrus in the Ground

In all areas of the Carolinas except for the extreme southern coastal areas, if you plan to grow citrus in the ground, you'll want to plant in a sheltered area on the south or west side of the house, preferably next to the house. The house will soak up heat during the day and release it at night, creating a little microclimate that's up to one zone warmer than the specified USDA zone for your area.

If planting near the house isn't an option, plant in full sun and allow enough room to put up a small pop-up greenhouse over the plants during the coldest nights of winter.

Citrus trees grow best in sandy loam soil, which means you'll want to amend the soil with compost and mulch the area after planting. Test the soil and add lime or sulfate to ensure that the pH is 6.0 to 7.0.

◾ *Pollination*

Most citrus can produce fruits without pollination. This is called parthenogenesis and is kind of like immaculate conception for fruit trees. Some tangerine trees require a cross-pollinator, but that will be indicated on the plant tag or in the product description of the website or catalog.

◾ *Maintenance*

Water: Citrus trees need more water when they're establishing themselves than after they've been in the ground for five or more years. Let the soil dry out between watering in-ground and container-grown citrus.

Fertilizer: The fertilizing schedule for citrus trees is slightly more complicated than for other fruit trees. Citrus require micronutrients in addition to macronutrients. Look for fertilizers specifically labeled for use on citrus. The fertilizers will probably have information about when to use. You can also follow the fertilizing schedule on page 107: (Note: This chart is for trees planted in the ground, not containers. In containers it is best to use a slowrelease fertilizer applied in spring and midsummer.)

Fertilizing Schedule for Citrus

Years since planting	Number of times fertilized each year	Pounds of nitrogen per tree, per year	Pounds of fertilizer per tree, per application		
			6-6-6	8-8-8	10-10-10
First year	6	0.15–0.30	0.4–0.8	0.3–0.6	0.3–0.5
Second year	5	0.30–0.60	1.0–2.0	0.8–1.5	0.6–1.2
Third year	4	0.45–0.90	1.9–3.8	1.4–2.8	1.1–2.3
Fourth year	3	0.80–1.0	4.4–5.6	3.3–4.2	2.7–3.3
Fifth year & older	3	1.1–1.4	6.1–7.8	4.6–5.8	3.7–4.7

Pruning: Citrus trees don't require a lot of pruning. If the tree gets to be too large or too full (so that there's not enough airflow and sunlight penetration in the center), you can prune in summer after the trees are finished flowering and fruiting.

Fruit thinning: Citrus trees will naturally drop about half of their flowers during bloom and then about one-third to half of their fruits before they reach maturity. That's a natural process and not something to worry about. If you want fewer, larger fruits, you can remove even more fruits before they develop by cutting them off the tree. There is a limit to the size any of the fruits will get though, and that is determined by the type of plant you are growing.

Pest Problems

If you grow potted citrus inside all winter (without leaving it outside during the warm days), you are likely to have problems with scale. My husband took his first lemon tree to work for the winter because he has big floor-to-ceiling windows. When the plant returned home in spring, it had the worst case of scale I've ever seen on a living plant. Because we are lazy gardeners most of the time, we did nothing to help the plant other than putting it back outside for the summer. The scale problem resolved itself when the plant was placed outside, because the plant became less stressed and natural predators attacked the insects. There are some sprays that will get rid of scale, too.

Harvesting

When are the fruits ripe? That's a good question with citrus. Most of the fruits will easily come off in your hand when they're ripe. Some trees have fruits that ripen over a period of weeks or months, so you can harvest for a fairly long time.

Most citrus trees bear fruit from late fall through early spring.

Blackberries ripening for harvest.

FRUIT SHRUBS & BRAMBLES

If you're looking to get into fruit growing, starting with shrubs is a good plan. Fruit shrubs are forgiving, easy to care for, and are prolific producers once established. Blueberries, figs, pomegranates, and blackberries are the easiest to grow in our areas. You could argue that pomegranates are trees, but in most of our growing area, they're best grown as small trees or shrubs rather than larger trees on the scale of, say, an apple or a pecan.

Landscaping with Fruit

Edible landscaping is a hot topic. Rosalind Creasy is arguably the pioneer of planting your yard with edibles. The first edition of her book *Edible Landscaping* was published in 1982 and reissued as a second edition in 2010. If you want to see what you can really do with edibles by converting your entire yard to a fruit and vegetable garden, her book is amazing.

Most people can't do that, though. But even if you're on the fence about turning your fence into a tomato trellis, you can at least tuck some Swiss chard or herbs into the perennial border.

It's easy to work fruit shrubs into the landscape. Blueberries are quite ornamental. They are one of the few plants that have reliable fall color in the Carolinas. Their leaves turn bright red—a plus for any garden. Fig trees make excellent screening plants, because they have large leaves and lots of them. (They are deciduous, though, so they are less useful as a screen in winter.) A fence of blackberries is somehow friendlier than an actual fence. I have a blackberry fence between my house and my neighbor's, and I get more berries than we can eat from just six plants.

Placing Fruit Shrubs in the Garden

Fruit shrubs can be planted throughout the yard (provided that you plant at least two or three near each other if they need to be cross-pollinated) or along the edges of the property. Because fruit shrubs take up less space and throw less shade than fruit trees, you can also plant them in or around your vegetable garden. Plant taller varieties on the north side of the garden so that they don't shade other plants.

Get kids involved with gardening by letting them help harvest!

BLUEBERRY *(Vaccinium* spp.*)*

Blueberries are one of the few widely grown fruits that are native to North America. Blueberries grow from Maine down to Georgia, over to Texas, and up to Wisconsin. Different varieties thrive in different locations. The key to success with blueberry growing is to choose a variety suited to your area.

■ *Recommended Varieties*

Rabbiteye blueberries grow well in the Piedmont and coastal areas of the Carolinas. They can grow to a height of 6 to 15 feet and can produce 8 to 25 pounds of fruit per bush after the third year. Cultivars include 'Brightwell', 'Tifblue', 'Climax', 'Premier', and 'Woodard'. Southern highbush is another type that does well in the heat, but it is more finicky about soil type. Try growing 'Misty', 'Sharpblue', 'Sunshine Blue', 'Legacy', and 'Star' varieties of Southern highbush.

■ *Planting*

Plant in full sun. Rabbiteye varieties will withstand some afternoon shade. Blueberries require cross-pollination, so plant at least two different varieties of each type of plant, 6 to 8 feet apart in the landscape, to ensure good pollination.

Blueberries require acidic soil. If your soil is alkaline, you will have to add aluminum sulfate to lower the pH to below 6.0. (If you can get the pH to 4.5 to 5.5, that's even better.) Take this step at least three months before you plan to plant. Blueberries also need relatively high levels of organic matter in the soil. If you have sandy soil, amend the soil in the entire blueberry planting area by incorporating at least 4 inches of a combination of compost and peat moss.

■ *Maintenance*

Water: In free-draining sandy soils, water every other day during the first six months after planting to encourage good establishment. After the first year, water twice a week during fruiting and once a week after plants have finished fruiting until fall, if it does not rain. Blueberries are shallow rooted, so they dry out quickly.

Blueberry plants provide fresh fruit in the summer and decorative foliage in cooler weather.

GROWING TIP

Establishment of moisture-retentive but well-drained, acidic soil is critical with blueberries. Fail to prepare your soil, and you will not get the results you want.

If the leaves turn yellow with green veins, the pH of the soil is most likely off and needs to be checked. Iron deficiency can be a problem in high-pH soils. Add chelated iron to correct the problem.

Mulch: Help keep the soil around blueberries moist by mulching with pine needles or pine bark mulch. Do not use hardwood mulch around blueberries as it can raise soil pH.

Fertilizer: Do not fertilize during the first six months of growth. After the first six months, feed with a slow-release fertilizer before the plants bloom and after you harvest the fruits, to encourage growth for next year. To maintain the acidity of the soil, you can also feed with fertilizer formulated for hollies. Blueberries are sensitive to fertilizer—be careful not to apply too much.

Pruning: Prune blueberries in late winter or early spring. They produce fruits on one-year-old growth, so don't hack the entire plant to the ground. If you do, you won't get any berries. You can prune a bit each year to keep the shrubs a manageable size. After trees reach five to six years of age, renewal prune by selectively removing one-third of the branches each year (throughout the shrub) to encourage new growth.

Pest Problems

Birds are the most annoying pest problem with blueberries. You can throw nets over the shrubs once the fruits are reaching harvestable size in order to protect them. Blueberry maggots and cherry and cranberry fruitworms can be problems with blueberries. You'll see the damage caused by these insects. Cranberry fruitworm make webs around the damaged berries. Consult your local Cooperative Extension for help controlling these insects.

Harvesting

Blueberry season runs from May through August in the Carolinas. Pick berries as you plan to eat them for the freshest fruits. (They can stay on the shrubs for a while, so you don't have to pick the shrub clean all at once.) The easiest way to determine if the berries are ripe is to taste a few. You'll learn to recognize when they're ready to harvest.

For best eating, pick figs when they are soft to the touch.

FIG *(Ficus carica)*

Figs are probably the most low-maintenance fruit plants you can grow. They grow best in the coastal and Piedmont areas of the Carolinas. It's a bit too cold for them in the mountains. It took a trip to California for me to truly appreciate what you can do with figs in the kitchen. My new favorite fall sandwich is an open-faced, toasted sandwich made with goat cheese, caramelized onions, and figs and braised in the oven for a few minutes. Figs can grow to be large shrubs, so give them plenty of room.

■ *Recommended Varieties*

Only grow varieties of figs that do not require cross-pollination. Recommended varieties include 'Celeste', 'Brown Turkey', 'Alma', and 'Kadota'. The variety 'Brunswick', also known as 'Magnolia', is best grown for preserves.

▨ *Planting*

Plant figs in full sun to partial shade in nutrient-rich, well-drained soil. Give them room to grow, as they can reach up to 30 feet in height and spread, though 10 to 15 feet is more common in the Carolinas.

▨ *Maintenance*

Water: Give figs 1 inch of water per week during summer.

Fertilizer: Fertilize once a year with a balanced, slow-release fertilizer in spring. In sandy soils, a second application of slow-release fertilizer in mid-summer is helpful.

Pruning: Figs need to be pruned mainly to keep them from taking over the landscape—they can get big. Prune figs in late winter while they are still dormant.

▨ *Pest Problems*

Figs are fairly pest-free plants, making them ideal for home gardens. The biggest problem affecting figs is root-knot nematodes. Inspect plants before planting them. If their roots look knobbly or knotty, do not plant them. Nematodes cannot be eliminated once established in the landscape.

▨ *Harvesting*

Figs taste best when harvested ripe. Pick figs when they are somewhat soft to the touch, and eat them within a few days.

GROWING TIPS

Because figs can experience pruning via freezing temperatures (frost damage), grow figs as shrubs, not trees. That way, if the plant dies back, you'll still have some of it left to bear fruit next year.

If you have a neighbor you'd rather not see, plant a fig tree along your lot line. They're dense shrubs.

POMEGRANATE *(Punica granatum)*

The pomegranate was most likely the "apple" referred to in the Bible as the fruit picked by Eve in the Garden of Eden. I'll never forget the first time I saw a pomegranate fruit cut in half. I was at the Apple Store in my hometown, manning the caramel apple station. There was a pomegranate half sitting in a bowl with another bowl covering it, but I didn't know that when I lifted the covering bowl. I thought something had gone horribly awry with an apple and we couldn't sell it!

Pomegranates have medium-thick red skin and lots of juice-covered seeds in membranes on the inside. Pomegranate juice is good for you. Eating the fruits can be interesting, as you have to either spit out the seeds or crunch them and swallow them. It's fun to grow pomegranates in the landscape though, because they produce beautiful orange-red flowers. The trees are ornamental edibles. Pomegranates produce fruit most reliably in zones 8–10 (primarily coastal and southern regions of the Carolinas), but they're worth planting for just their ornamental value.

■ Recommended Varieties

There are tall or small treelike cultivars and dwarf cultivars.

Taller Than 6 Feet: 'Wonderful', 'Double Red', 'Early Wonderful', 'Flavescens' (yellow flowers), 'Granada', 'Eight Ball' (black fruit)

Shorter Than 6 Feet: 'State Fair' (shorter cultivar), 'Nana' (dwarf)

■ Planting

Plant pomegranates in full sun in well-drained soil. If you live in the cooler regions of the Carolinas, plant in a sheltered location to prevent cold damage. Planting at least two plants also improves the chance of fruit set.

■ Maintenance

Water: Pomegranates are fairly maintenance-free plants. Water to establish the plants, and then water during summer only if your region is experiencing a drought.

Fertilizer: Fertilize in March and July with a slow-release fertilizer to encourage fruiting.

■ Pest Problems

Pomegranates are mostly pest-free.

GROWING TIP

If you get poor fruit set, it's probably because of poor pollination or high humidity. Pomegranates rarely set heavy fruit loads in the Carolinas.

◼ *Harvesting*

Pomegranates are usually ready to harvest from August through October in the Carolinas. The fruits might not be as large as the fruits you purchase at the grocery store. Look for fruits that are the size of a tennis ball and pick them individually. Store in the refrigerator until they can be eaten.

Pomegranates produce the best fruit in the warmest areas of the Carolinas.

BLACKBERRY *(Rubus* spp.)

Blackberries are easy to grow throughout the Carolinas.

Blackberries are "bramble" fruits. Brambles are part of the rose family. In fact, their flowers look a little bit like small wild roses. Blackberries and raspberries require similar care and have similar growth habits, but only blackberries will grow in all parts of the Carolinas. Fluctuations in winter temperature and extreme summer heat make growing raspberries an exercise in frustration in coastal areas, though raspberries can be grown in the western Piedmont and mountains. Blackberries, however, are easy to grow throughout the Carolinas and have few pest problems. A half-dozen plants will give you a large harvest, as well.

■ Recommended Varieties

There are semitrailing and erect types of blackberries that can be either thorny or thornless. Semitrailing varieties of blackberries are well suited to the warmest areas of the Carolinas. Erect types grow well throughout but are slightly more cold hardy, making them the best varieties for the mountains.

'Black Satin' and 'Hull' are semitrailing types suitable to all growing areas in the Carolinas. 'Gem' is a thorny, semitrailing variety that is resistant to rosette disease.

'Cherokee' is an erect type that produces a good crop of fruit but is susceptible to rosette disease (which might not be a problem in your area). 'Navaho' is similar but has better resistance to rosette disease.

You can purchase both container-grown and bare-root plants. If container-grown plants are available in your area, spend the extra money for them. They'll establish themselves better and faster.

■ Planting

Plant blackberries in full sun in well-draining soil with a pH of 5.5 to 6.5. Sun is important for growth and fruit production. Some fruits will produce in partial shade, but not blackberries.

Trellising Systems for Trailing and Erect Blackberry Types

(A) Train trailing plants to a two-wire trellis. (B) Train erect blackberry plants to a one-wire trellis. Note that trailing blackberries need more space between plants than erect blackberries.

You can plant container-grown plants at any point in the year, as long as snow isn't on the ground and the ground isn't frozen. Plant bare-root plants in the winter while they are dormant. Soak the roots of bare-root plants for twenty-four hours before planting. Follow instructions on page 46 for planting bare-root plants.

Plant container-grown plants so that the rootball of the plant is at the same level as, or just slightly higher than, the soil around the planting hole. Don't plant too deep!

Brambles grow best with support. You could stake individual plants or grow them against a fence. The best way to grow them though, is to set up a post and wire system with 2 inch × 4 foot posts and heavy-duty wire.

You can train the plants to grow along the wire, and you can anchor the stems using coated tomato wire or twist ties. (I leave several twist ties on the wires so that there's one handy if I happen to decide to go outside and check the blackberries. At the end of the season, I cut the old canes and refasten the twist ties so that they're handy again in spring.)

■ *Maintenance*

Water: Frequent watering during the first year is the best way to make sure the plants get off to a good start. Run soaker hoses along the plants (taking care to keep the hoses from touching the stems). Water every two or three days during the first growing season in sandy soils. In subsequent years, watering at least twice a week while the fruits are swelling, giving the plants 1 inch of water per week, will help the plants produce a big crop of fruits for harvest.

Fertilizer: Fertilize in spring with a slow-release fertilizer.

Pruning: Improper pruning of blackberries resulting in no fruit production has broken many a heart. Blackberry stems are called canes. A first-year cane is called a primocane. A second-year cane is called a floricane.

You won't hurt fruit production in your plants if you follow this rule: only prune blackberries all the way to the ground in fall, and only cut down canes that bore fruit during that year. If you only prune the canes that bore fruit, and you only prune in fall, you won't cut down fruit-bearing canes before their time.

The floricanes of summer-bearing blackberries usually start to die back after you pick the fruits anyway, which makes it easy to see which ones to cut down.

Another type of pruning is pruning to produce more productive side shoots. Prune blackberries back by 6 inches in the middle of summer. This will encourage the plants to produce more side shoots off the main canes, which will result in a bigger fruit crop during the second year.

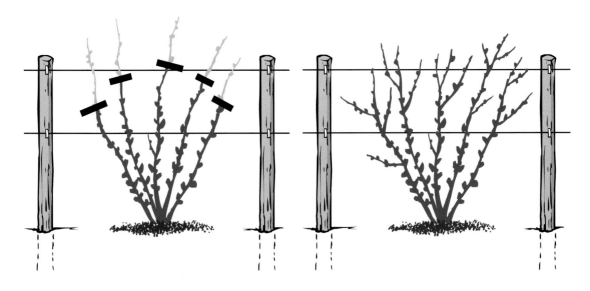

If you cut off the top 6 to 12 inches of the canes during the first year, the plants will produce side branches that bear fruit during the second year.

Pest Problems

Blackberries have few pests that are a major problem in the home garden. There are a few diseases that affect plants, including rosette disease. Plant resistant varieties to avoid this problem.

Cane borers attack blackberries. You will be able to tell if you have these insects if the top of a first-year cane wilts and falls over suddenly. Cut below the wilted area and discard the branch with the borer. (Do not compost.)

Occasionally birds will eat your blackberries before you can. If this is a problem, purchase bird netting at a garden center or home-improvement store, and throw the nets over the plants.

Harvesting

Harvest blackberries when they easily come off the plants. Plants will produce ripe berries for several weeks. Harvest only what you want to eat, and harvest fresh daily.

Raspberries can be grown in the cooler areas of the Carolinas.

Muscadine grapes have thick skins.

VINING & GARDEN FRUITS

After covering fruit and nut trees, citrus, and shrubby fruits, what's left? Vining fruits and garden fruits. Vining fruits such as muscadine grapes and hardy kiwis are perennial fruits with sprawling growth habits. Melons are similarly sprawling, but are annuals, though most gardeners prefer to grow them as perennials. Strawberries are somewhat in a category of their own—they can be grown as perennials or annuals. Melons and strawberries could be covered in the "vegetables" section, but they're sweet, and we treat them as fruits in the kitchen.

This category of fruits includes plants that can be somewhat tricky to site in the garden. They all take up a fair amount of space, and all but melons are perennial. The question to ask yourself before deciding to grow any of these fruits is, how badly do you want your own?

To get enough strawberries to make a difference on your table, you have to plant a lot of plants, and that takes space. Hardy kiwi is more of a novelty than anything else, though its fruit does taste good. Melons take up a huge amount of space. I'm not trying to talk you out of growing these, but when it comes to making a real difference in the budget versus taking up a lot of valuable real estate in the garden, these fruits are hogs. Strawberries and melons are cheap to get at the farmers' market when they're in season. If you are low on space, consider growing a few plants for fun but leaving the heavy lifting to market growers in your area.

The main vining fruit that I'd recommend making room for is the grape. Muscadine grapes just taste like a grape should taste. They might be novelties, but they're tasty novelties. You can also make some pretty tasty juices and cocktails with them, and they're hard to find in the stores—and expensive.

Fresh strawberries have the best taste, but can they can take up a lot of room in the garden.

Hardy kiwi is a bit of a novelty, but the fruit does taste good.

HARDY KIWI *(Actinidia arguta)*

Hardy kiwi plants are in the kiwi family, along with the fuzzy brown fruits you buy at the grocery store. Hardy kiwi fruits are the size of kumquats or large grapes, and they have edible skin. The reason you don't see these fruits for sale in the stores is that they taste best when harvested completely ripe, and they don't ship well. They're excellent for the home fruit garden, though. If you have the space and the patience, they're fun to grow.

■ *Recommended Varieties*

Grow the straight species kiwi (*Actinidia deliciosa*) in zones 8 or higher. In zones 7 and above, plant *Actinidia arguta* and *Actinidia kolomikta* (Russian kiwi). Finding a named cultivar is less important than making sure that the species (second part of the botanical name) is correct.

Kiwis grow as male or female plants. The females produce the fruits, but males are needed for cross-pollination. You need at least one male plant for every nine female plants. (Nine hardy kiwi plants would produce a lot of fruit. The home gardener probably only needs two or three female plants.)

Planting

Kiwis are big vines that grow as much as 20 feet in a year. Plant them next to a sturdy fence, or put up a trellis made from metal pipe. They need full sun to partial shade and a soil pH of 5.0 to 6.5 (slightly acidic). Plant vines 10 feet apart to give them space to grow.

Maintenance

Water: Water once a week in the summer after the first year.

Fertilizer: Fertilize in spring with a slow-release fertilizer.

Pruning: The biggest issue with hardy kiwi is that the vines are large. You can prune them somewhat to control size, but don't plant these if you don't want them to sprawl. The plants attach to trellises by twining around them, so if you're growing them in front of a wall or fence, you'll have to tie the vines to the fence.

Pest Problems

Hardy kiwis have few pest problems.

Harvesting

It takes hardy kiwis up to five years to bear fruit. Fruits ripen in late summer to early fall and should be picked when they are just soft to the touch. The skin is edible.

A hardy kiwi vine showing fall color.

MUSCADINE GRAPE *(Vitis rotundifolia)*

To me, muscadine grapes taste just the way a grape should taste: a little sweet, a little tart, and full of flavor. Sometimes called scuppernongs, muscadines are mostly known for the ridiculously sweet wine made from them. I'm not a big fan of the sweet wine, but I enjoy eating the fresh grapes. They have tough skins and big seeds, but I'd rather suck on a scuppernong than a piece of grape-flavored hard candy any day.

■ *Recommended Varieties*

You can grow European and American bunch grapes (which are the "table grapes" you buy at the grocery store) in the mountains and western Piedmont, but they aren't well suited to coastal regions. Look for muscadine grapes when shopping for grapevines for the coast and Piedmont.

Improved cultivars that are more disease and pest resistant than earlier types are 'Carlos', 'Doreen', 'Magnolia', 'Nesbitt', 'Noble', 'Regale', and 'Triumph'.

These varieties also have "perfect flowers," which means they have male and female parts, don't require more than one vine to produce fruit, and don't require cross-pollination from another plant. Check the plant tag or catalog description when buying grapes to make sure they don't require cross-pollination. Some grape varieties have only female flowers. Those require at least one plant with perfect flowers that bloom at the same time in order to produce fruit.

You can buy bare-root grapes and container-grown grapes. Plant bare-root grapes in February to late March. Plant container-grown grapes at any point during the year.

■ *Planting*

Plant grapes in full sun in well-drained. Grapes do not grow well in heavy clay soils or soils where water sits for more than a few hours at a time.

Grapes require support. If you don't want to start over every couple of years, it is best to go big or go home. A post and wire system like you'd use for blackberries, but on a larger scale, it will work well for grapes. Situate the wire at a height that's comfortable for you to reach—between waist and chest high works well. Plant grapes with 10 to 20 feet of space between each vine.

■ *Maintenance*

Water: Water grapes two to three times a week as they are becoming established, more often in sandy soils. In subsequent years, give vines 1 inch of water per week during the fruiting season.

Fertilizer: Muscadine grapes benefit from the application of 2 ounces of ammonium nitrate per vine, spread on the ground in a 2-to-3-foot-diameter area around the trunk. Fertilize in March, May, and July. If vines are growing too vigorously, skip the July application or just apply slow-release fertilizer in March.

Muscadines are sensitive to magnesium levels in the soil. If you see yellowing in the spaces between veins of older leaves, there's probably a magnesium shortage, or the pH is too high. Spread Epsom salts in the same diameter area you'd spread fertilizer at a rate of 2 to 4 ounces per plant to correct this problem.

Baby muscadine grapes on the vine.

Pruning: The biggest mystery of grape production is pruning. Grape vines have a central stem, called the trunk, and two side shoots called cordons. The plants will bear fruit from the cordons on compressed side shoots called spurs. Each cordon will have many spurs. From the spurs grow shoots with buds that will sprout and produce grapes.

Train new grape vines by allowing them to grow up toward the wire of the trellis. Cut off side shoots while the vine is growing up. Once the vine has reached the wire, cut off the growing tip to force side shoots to grow. Two of these will become the right and left arms of the cordon.

Prune grapes in winter. At each spur, prune back all growths except for one piece of fruiting spur from the previous year's growth. Each fruiting spur should be about 4 inches long and have three to five buds.

To keep new fruiting spurs forming, each year remove every other spur down to the cordon. New spurs will sprout.

It's important to know that muscadine grapes bear fruit on growth that arises from buds on one-year-old spurs, so you don't want to cut off that growth when pruning.

■ Pest Problems

Muscadine grapes are relatively pest-free.

■ Harvesting

Ripe grapes will easily come off in your hand. You can cut individual clusters as they ripen. Do a taste test to see if the grapes are ready to eat!

Melons can take up a lot of space in the garden, but they're fun to grow.

MELON Watermelon *(Citrullus lanatus);* Cantaloupe and Honeydew *(Cucumis melo)*

Melons are annual vines related to cucumbers and squash. If you're limited on space in the vegetable garden, skip the melons and buy them at the farmers' market. If you have space, give them a shot. It's fun to go out into the garden and find tiny watermelons growing larger each day.

■ *Recommended Varieties*

Watermelon: 'Charleston Gray', 'Crimson Sweet', 'Golden Crown', 'Royal Sweet', 'Tiger Baby'

Cantaloupe: 'Burpee Hybrid', 'Ambrosia', 'Park's Whopper', 'Athena', 'Scoop II'

Honeydew: 'Earlidew'

■ *Planting*

Direct-sow melon seeds in the soil when soil temperatures are at least 65 degrees F. It's important to plant melons when the soil is warm enough, but also to allow enough time for the plants to flower and fruit before air temperatures routinely reach above 80 degrees F during the day for cantaloupe and honeydew and 85 degrees F for watermelon. That can be a short window in the Carolinas.

Sow seeds outside in the southernmost areas of the Carolinas after all danger of frost has passed. This can be as early as mid-March, continuing to late May in the mountains. Use a soil thermometer to determine when it's time to plant.

Melons need full sun and moisture-retentive but well-drained in order to produce well. The soil pH should be 6.5 to 7.0, though watermelons can tolerate a pH as low as 5.5. Plant seeds in hills at least 4 feet apart, planting three seeds per hill. Thin plants to one or two plants per hill after they grow three sets of leaves.

Maintenance

Water: Melons are heavy drinkers. Lay soaker hoses after the plants have sprouted, and water every other day during the growing season in sandy soils. Avoiding overhead watering can prevent some disease problems.

Fertilizer: Sidedress with calcium nitrate when the vines are 3 feet long, or apply a slow-release fertilizer at planting time.

Pollination: Melon plants have separate male and female flowers on the same plant. The flowers have to be pollinated in order to produce fruits. If you want to be absolutely sure that plants have been pollinated, you can hand-pollinate them. Move pollen from a male flower to a female flower using a paintbrush or cotton swab. (Female flowers have a small swelling behind the flower, which is what will grow into the melon.)

Pest Problems

Melons suffer from the same pests that other squash-family plants encounter. Flea beetles and squash bugs are some of the worst but rarely affect melons as severely as squash.

Harvesting

When are melons ripe? That is the hardest thing to learn, and something you'll figure out over time. Cantaloupe varieties will be ready to harvest around 30 to 35 days after flowering; they should smell sweet. Honeydews take about 40 to 45 days after flowering; they turn pale green.

Watermelons take longer, but the time to maturity differs greatly, depending on the type of melon. Check the seed packet for information about the number of days from planting to harvest or from flowering to harvest. You can tell if watermelons are ripe by looking at the tendril closest to the fruit. When that turns brown, the melons are ready to pick.

STRAWBERRY *(Fragaria × ananassa)*

Warm, fully ripe, fresh-picked strawberries are one of the most delicious things you can eat. Strawberries can be grown as annuals—replanted each fall for fruiting in spring. This is the way commercial farms grow strawberries. Most gardeners prefer to grow strawberries as perennials via the mat and row system. If your plants grow well when left in the ground for several years, it is absolutely easier to treat them as perennials this way. If anthracnose is a problem and plants decline or don't fruit, switch to the annual system of fall planting and spring harvesting.

■ *Recommended Varieties*

There are short-day (June-bearing) strawberries and day-neutral (everbearing) strawberries. Short-day varieties of strawberries do best in the Carolinas. As with many fruit crops, choose your variety carefully. If you have u-pick farms in your area, check them out. They will often have plants for sale.

Because of the number of diseases that affect strawberries in the Carolinas, buy resistant varieties when possible.

Recommended varieties for the Carolinas are 'Florida 90', 'Sunrise', 'Earliglow', 'Cardinal', 'Surecrop', 'Sweet Charlie', 'Apollo', 'Delite', 'Chandler', and 'Douglas'.

Use nets to ensure you'll get to eat your strawberries before the birds do.

■ *Planting*

Strawberries have an unusual growth cycle. For success growing them at home, you need to understand the growth cycle and plant accordingly.

Strawberry plants produce runners during summer. In fall, the plants stop spreading out and flower buds form in the crown of the plant (the center of the plant—a compressed stem). In early spring, the crown sprouts leaves and flowers. Berries form four to five weeks after the flowers open. It takes about three weeks for berries to ripen. When the days grow longer and the temperature warms, plants send out runners again.

Strawberries grow best in full sun in well-drained, sandy soils with a pH of 5.5 to 6.5. They do not tolerate heavy clay soils or swampy, poorly drained soils very well.

To grow plants as perennials, plant strawberries 2 feet apart with 4 feet between rows. Allow runners to fill in the row to create a 2-foot mat with a final spacing of 2 feet between rows.

You can plant new strawberries in spring and allow them to send out runners throughout summer that will bloom and fruit the following year. You can also transplant larger plants in fall and allow them to produce fruits in spring.

If anthracnose is a problem, dig up plants after harvesting, and plant new plants each fall in a new area.

Maintenance

Water: Strawberries require extra water at very specific times during their growth cycle:

- Right after planting
- During peak harvest
- After peak harvest when you're encouraging new runners
- In fall when fruit buds for the following year are forming

During these times, water the plants so that the soil is wet to a depth of 6 to 8 inches.

Mulch: All strawberries do best when mulched to conserve moisture, keep weeds under control, and keep the berries from touching the soil.

Fertilizer: Apply slow-release fertilizer at planting time or in spring and again in early summer. Apply a granular fertilizer in September.

Pest Problems

There are numerous pests and diseases that affect strawberries. Birds are the worst animal pests. Use nets to keep the birds from eating the fruits as they ripen. Multiple fungal and bacterial problems attack strawberry plants. If you think your plants could be afflicted, the best thing to do is to dig up a plant and take it to your local Cooperative Extension office for help with diagnosis and treatment solutions.

Harvesting

Harvest when the fruits are colorful and can be easily pulled off the plant.

CHAPTER 8

VEGETABLE & HERB GARDENING

The fact that homegrown tomatoes taste better than store-bought tomatoes is not only a cliché—it's true. Why? Because when you can pick your produce right before you use it, as opposed to a month before it's going to be used, you can let the fruits develop to their full ripeness and sweetness. You can also grow more delicate varieties that don't have to withstand shipping and warehousing.

Cool-Season versus Warm-Season Vegetable Gardening

If you grew up in the Carolinas, you're probably used to two seasons of growing vegetables. If you're a transplant from colder northern climes, the fact that you can garden year-round in most areas of our states will be a revelation. It will also cause you no end of frustration unless you spend some time learning about cool-season versus warm-season gardening.

Growing herbs in a small kitchen garden makes it more likely that you'll use them.

Certain plants are absolutely happier and healthier growing from fall to spring. Others need hot weather and only thrive during summer. There are a few vegetables that will, if planted, continue to grow through summer if planted early enough in spring. The key is planting the right plant at the right time of year so that you get the biggest harvest and are successful the first time around.

The vegetable and herb profiles in this book are organized to help you distinguish cool- and warm-season vegetables so there is no question in your mind as to what you should be planting when.

Transitioning between Warm- and Cool-Season Gardening

It's easier to transition a vegetable garden between seasons if it is planted in rows. If you plant in rows, you can easily save part of the garden for the next season. Or you can neatly pull up all of one crop and plant another.

Cool season vegetable and herb garden

Raised beds and small-space gardens can be more challenging. You have to find ways to shoehorn in the new crop while the old crop is finishing up. If you're like me, and you plant the same kind of vegetable in different parts of the garden (I do it partially to thwart pests and partially because I'm disorganized), you can end up with small spaces all over the garden. You'll find yourself wandering around trying to decide what to plant where, not wanting to leave any of the valuable and limited space empty.

A little bit of planning goes a long way when making the transition. If you have the time, patience, and organizational skills to do this, sit down with some pieces of paper and make a grid that shows your garden. With a colored pencil, write in the cool-season plants you plan to grow. In a different-colored pencil, write the warm-season plants that you're planning to plant when the cool-season vegetables are finished. This small bit of organization will also help you keep yourself from going crazy when you go plant shopping—you'll have a better idea of what is going in which space.

Small-Space Vegetable Gardening

Don't say no to edibles, even if you have only a small amount of space in the sun. It's easier than ever to grow enough vegetables to supplement your diet. Raised-bed kits, self-watering containers, and vertical-gardening assemblies allow you to grow vegetables anywhere, with any amount of space. In the Carolinas, growing in raised beds or containers lets you more easily control the quality of the soil. (We have notoriously bad soil.)

Find Your Favorite Vegetables and Herbs

Ready to get growing? Use this mini-index to find the plants you want to learn about.

Cover Cropping
If you don't plant the entire garden with cool-season vegetables, consider cover cropping. This is the practice of planting a nitrogen-fixing legume such as fava beans, clover, hairy vetch, or Austrian winter peas in a section of the garden that you're not using. You will let these plants grow until they are about to flower. Then, cut them down and turn them under into the soil as a "green manure." Cover cropping is an inexpensive way to add nutrients into the soil while preventing erosion from wind or water during winter.

Crop Rotation
Certain plants in the cool-season vegetable garden are highly susceptible to soilborne diseases. Changing where you plant these plants from season to season helps avoid those problems. This isn't always an option with smaller gardens, though. Cabbage-family plants, including broccoli, cauliflower, collards, kale, and cabbage, are the ones that have the most problems with soil diseases and fungi. If you find yourself facing major problems with diseases of the soil, consider growing just those crops in containers with new soil each season.

You can see the rows of newly planted seeds delineated by the lighter-colored seed-starting mix covering the seeds.

Cool-Season Gardening Tips & Tricks

The cool season of vegetable gardening runs from roughly September through April in all but the mountainous regions of the Carolinas. In the mountains, the cool-season garden grows from March through May. The confusing part of cool-season gardening is that there are two parts to it: October through January or early February, and February through April. In a way, it's almost like we experience three gardening seasons. If viewed from a strictly calendar-oriented perspective, the gardening year in the Carolinas goes like this:

Cool season 1 (January to April)
Warm season (April to September)
Cool season 2 (September to December)

Some plants, when planted in fall, can be harvested through spring. Others require two different plantings in order to get a continuous harvest. This is noted in individual plant descriptions where applicable.

To heap even more confusion onto the plate, if you want to grow some cool-season vegetables from seeds sown directly into the garden in fall, you will actually sow the seeds in August or September. But wait, aren't those months part of the warm season?

Welcome to vegetable gardening in the Carolinas. It never stops.

Certain plants can sprout from seed when soil temperatures are relatively high (75 to 85 degrees F) but will grow best when air temperatures start to cool down. Broccoli, cabbage, kale, and many other members of the cabbage family are like this. You will sow these plants outside in late summer, but they will grow and mature during the true cool season. If you grow these plants from transplants, you'll plant out later in the fall or early in spring to take advantage of cool weather.

Plants listed in the cool-season section of this book grow best and mature during cooler parts of the year. Most won't grow, other than to germinate, when the weather is too warm. Most of the vegetables in the cool-season vegetable garden palette are originally from northern sections of eastern Europe and thus are well adapted to cool, humid weather. (That sounds like a Carolina winter!)

Planting Seeds

Many of the cool-season vegetables are easy to grow from seeds planted directly into the garden. This is indicated in the individual plant profiles.

Sowing Seeds Outside

Follow the instructions on the seed packets and in the plant profiles regarding the planting depth. Some seeds, such as peas, need to be planted fairly deep, while others, such as lettuce, barely need covering.

Cover seeds with seed-starting mix instead of regular garden soil. This lightweight mix makes it easy for sprouting seedlings to break out above the soil.

Keep seeds moist while they are sprouting. Established plants can deal with a bit of drought or water fluctuations, but seeds must stay evenly moist while sprouting.

Planting Transplants

Some plants are much easier to grow from transplants—particularly if you're planting a second round of vegetables in February or March. The cooler soil temperatures of late winter and early spring cause seeds to take longer to germinate. During that time, they're more susceptible to fungal problems, particularly root rot diseases. Whether you start your seeds inside and grow your own transplants or you purchase transplants, follow these tips for success:

Working with Transplants

Always harden off transplants before planting outside—even if you bought them at a garden center. Hardening off is the process of getting plants used to outdoor conditions, including sun, temperature, and wind. Start by putting your transplants on a porch or sheltered location for about a week. Then move them to a less protected area for another week. Finally, plant in the garden bed.

Plant transplants at the recommended finished spacing for the mature plants. Transplants are expensive! You don't want to lose money by having to "thin" transplants later.

Planning for a Continual Harvest

Succession planting is the process of planting the same plant, several days apart, several times throughout the growing season. Lettuce, carrots, radishes, and turnips are examples of plants that you can plant in succession. These vegetables are quick to mature. By planting several crops of these vegetables on a staggered schedule, you ensure a bountiful harvest of fresh produce throughout the growing season. If a vegetable is a good candidate for succession planting, this is indicated in the plant profile.

Planting carrots every few weeks ensures a continual harvest.

Harvesting

If you plant carrots in fall, you can harvest them throughout the entire winter. The garden bed acts as a sort of "cold storage." Soil conditions during winter are much like a refrigerator: cool with medium humidity. If you want smaller carrots, planting a few successive rows will keep you going for longer, and you can basically use the garden bed as a "storage area" for the harvest. This also works for parsnips.

Other vegetables can be harvested at any point during their growth cycle. Turnips and beets, for example, taste just as good when harvested at 1 inch in diameter as they do when harvested at 3 inches. You can't leave them in the ground, but by planting successive rounds of seeds, you can harvest all winter, at various stages of the growth cycle.

Still other plants have to be harvested when they're ready—there's no early harvesting or "cold storage." Cabbage, broccoli, and cauliflower are done when they're done and need to be picked and eaten.

Over time, you will get the hang of which vegetables can stay in the garden until you're ready for them, which ones can be eaten early, and which ones you need to harvest at a specific point in the growth cycle.

ARUGULA *(Eruca sativa)*

Arugula is an annual green that thrives during cool weather but that will also survive our warm and humid summers as long as it is watered regularly. Summer-grown arugula has a wasabi-hot flavor—different than the milder winter greens. Arugula is hardy to temperatures around 20 degrees F, which means you need to have some frost cloth handy if the forecast is for colder weather. Once the plants start to bolt (send up flower stalks), they're nearing the end of their life cycle. You can cut off the stalks to prolong leaf production, but eventually it's better to move on to harvesting from newer plantings.

■ Recommended Varieties

The straight species of salad arugula grows fine in the Carolinas. The salad arugula variety 'Astro' is heat tolerant, a plus for our region. Wild arugula (*Diplotaxis tenuifolia*) is a type of arugula with longer, more deeply incised (cut) leaves. It is spicier than salad arugula.

■ When and Where to Plant

To enjoy fresh arugula all year, you'll need to sow successive plantings every two weeks. Seeds planted in May will yield harvests throughout the summer.

Temperature: Arugula grows best in cool weather but will grow all year in all Carolinas regions except for the mountains. Sow seeds outside from August to October, and February to May.

Soil: Plants thrive in deep, fertile soil with lots of organic matter. The best pH is 6.0 to 6.8.

Sun: Full sun to partial shade.

■ How to Plant

Starting seeds indoors: You can start seeds indoors year-round if you prefer working with transplants in the garden bed. You do not need a heat mat, as arugula prefers cooler temperatures.

Planting outside: Sow seeds outside from August through October and February through May.

■ How to Grow

Water: Arugula needs steady moisture, particularly as the weather warms, to keep it from bolting.

Fertilizer: Sidedress with a balanced fertilizer (10-10-10, for example) once a month to encourage steady leaf production, or apply a slow-release fertilizer at planting time.

GROWING TIPS

Gardeners in the mountains can plant arugula outside in March for a spring/summer harvest and in August for a fall harvest.

For milder leaves in summer, provide arugula plants with some shade.

To keep pests away from arugula, plant away from cabbage-family plants (which harbor the same pests).

Pest control: Flea beetles are the most common pest to strike arugula. Use floating row covers to control these pests when they are active.

When and How to Harvest

Arugula is cut and come again. Young leaves are tender and mild, while older leaves are spicier and better for braising. Use scissors to cut the leaves at the base of the plant. Do not pull up the entire plant.

Mix arugula in with other salad greens.

BEET *(Beta vulgaris)*

Beets are primarily a root crop, but their tops—or leaves, which are also called beet greens—are tasty and nutritious, too. Young greens (or tops) are great in salads, while the larger greens you get when harvesting beets taste good when braised and tossed with pasta. Beets need full sun and loose soil to form good roots. As with all root crops, beets grow best when sown directly into the garden, rather than planted as transplants. If your only experience with beets is pickled varieties on the salad bar, you're missing out. Beets taste delicious when roasted. You can also grate raw beets onto your salads. In addition to red varieties, you can also grow white, yellow, and striped varieties.

■ *Recommended Varieties*

The following red varieties grow well in the Carolinas: 'Ruby Queen', 'Early Wonder', 'Red Ace', 'Pacemaker II', and 'Detroit Dark Red'. 'Golden' is a yellow variety. 'Chioggia' is a red-and-white-striped variety.

■ *When and Where to Plant*

Temperature: Beets germinate best when soil temperatures are 55 to 75 degrees F. The plants grow best when air temperatures are 60 to 65 degrees F. In the mountains, sow seeds outside April 1 to 30 and again August 1 to 31. In the Piedmont, sow seeds outside March 1 to 30 and

Both the tops and the roots of beets can be eaten.

again August 1 to 31. In coastal regions, sow seeds outside from August 1 to September 15 and again from February 1 to May 1.

Soil: Beets need deep, loose, fertile soils with a pH of 6.0 to 7.0.

Sun: Full sun

■ *How to Plant*

Starting seeds indoors: Not recommended

Planting outside: Sow seeds outside 2 inches apart and ½ inch deep.

■ *How to Grow*

Water: Beets need even moisture to avoid scab, a condition in which brown, raised patches form on the outsides of the roots.

Fertilizer: Use a balanced fertilizer when planting, and follow up with a low-nitrogen, high-phosphorus fertilizer when the tops are 4 inches tall. Only apply phosphorus if a soil test indicates it is needed.

Pest control: Cercospora leaf spot is a fungus that causes brown scabby spots on leaves and interferes with sugar production and root development. This can be treated with fungicides.

■ *When and How to Harvest*

To harvest beet greens for salads while keeping the roots growing, use scissors to cut no more than one-quarter of the leaves at any time. To grow beets for this dual purpose, it is best to use small beet greens only as an addition to salads, not as the only green in a salad bowl. To check if roots are ready to harvest, pull one up as a test. Remember: plants you grow at home could be different varieties or have had different care than those you see at the grocery store. Just because your beets might be smaller than you expect doesn't mean they aren't ready to eat. Once beets reach their indicated maturity (usually after 60 days growing), it's best to harvest. If you leave them in the ground, they'll become woody and unpleasant to eat.

GROWING TIPS

Beets will develop internal black spot if boron isn't available in the soil. This is usually because your soil pH is too high or too low. If this is a problem in your garden, apply borax at a rate of 1 pound per acre at least seven days prior to planting and adjust soil pH according to soil test recommendations.

Beets planted at the proper time have fewer pest problems than beets planted late. If you're confused about when to plant, err on the side of earlier.

Newer hybrid varieties are sweeter than older varieties.

BROCCOLI *(Brassica oleracea,* Italica Group*)*

Broccoli is the go-to vegetable for the dinner table or the picnic basket. Like all vegetables in the cabbage family, broccoli is highly nutritious because it is filled with vitamins and minerals. Research studies have also shown that broccoli has anticancer properties—when eaten raw. In most of the Carolinas, we have the opportunity to grow multiple crops of broccoli throughout the cool season. (In the mountains, broccoli is best grown as a spring crop.) Change where you plant broccoli in the garden from year to year to prevent plants from succumbing to soilborne diseases.

■ Recommended Varieties

'Packman' and 'Southern Comet' are two varieties that grow best during the early spring season. Both of these varieties produce side shoots after the main head is harvested.

For disease resistance, including head rot and downy mildew, plant the 'Everest' cultivar. Other suggested varieties include 'DeCicco', 'Premium Crop', 'Green Duke', and 'Emperor'.

■ When and Where to Plant

Temperature: Broccoli germinates best when soil temperatures are 50 to 85 degrees F. It grows best when temperatures are in the 60s F. (This means you can direct-sow a fall crop, but not a spring crop outside.) For a spring crop, plant transplants outside from February 15 to March 1 in coastal regions and March 1 to March 30 in Piedmont or mountain regions. For a fall crop, plant outside from August 10 to September 1 in coastal regions and July 15 to August 15 in Piedmont and mountain regions.

Soil: Soil composition is less important (broccoli will grow equally well in sandy soils or clay-filled soils) than soil pH. The soil pH for broccoli must be 6.0 to 6.5. If the soil pH is lower than 6.0, mix lime into the soil at least two months before planting.

Sun: Full sun

Successive crops of broccoli can be grown throughout the cool season in the Carolinas.

GROWING TIP

Some varieties of broccoli are extremely temperature sensitive. If there's a cold snap (temperatures of 35 to 50 degrees F) for more than ten days after broccoli is planted, the plants will bolt (flower) and you won't get a head to harvest. You can try to prevent the cold from affecting the plants by covering them with a cold frame. Once the plants bolt, you have to just pull them up and start over. This is more of an issue in the spring. Broccoli plants grown in fall rarely have problems with bolting.

■ *How to Plant*

Starting seeds indoors: Start seeds indoors at least 40 days before you want to plant transplants outside. Broccoli takes 3 to 10 days to germinate and about a month to grow to transplantable size.

Planting outside: Always harden off your transplants before planting them outside in spring. It is easier to direct-sow a fall crop and get good results than it is to direct-sow a spring crop outside. Go with transplants for spring. Leave 12 to 18 inches between transplants, or thin seedlings to this spacing after direct-seeding.

■ *How to Grow*

Water: Keep broccoli evenly moist throughout the growing season.

Fertilizer: Add compost to the soil before planting. Then sidedress with calcium nitrate when the plants are 6 inches tall.

Pest control: Broccoli and other cabbage-family plants are susceptible to a variety of pests, including aphids, cabbage whites, cabbage loopers, diamondback moths, flea beetles, harlequin bugs, and slugs. Plant resistant varieties when possible, and use floating row covers when flying insects are active. Use Neem oil and *Bacillus thuringiensis* (B.t.) to control a wide range of pests.

■ *When and How to Harvest*

Cut the center flower head when it is still dark green and tight (about 60 days for most varieties—it will say on the seed packet). When the head starts to loosen and turn yellow, it's too late to harvest. If you're growing a variety that produces side shoots, leave the rest of the plant in the garden for up to two months. If the variety is primarily grown for the large center head, cut the heads and pull up the rest of the plant to make room for other vegetable crops.

BRUSSELS SPROUTS
(*Brassica oleracea,* Gemmifera Group)

Brussels sprouts—you love them or you hate them. I love them sautéed with butter and bacon, but that completely defeats the purpose of eating a vegetable. One reason to grow this unusual-looking cabbage-family plant is that it's easy to grow if you can keep the pests away and are patient. Leave Brussels sprouts in the garden until there's a frost, which will make the "sprouts," which look like mini-cabbages, sweeter.

The "sprouts" of Brussels sprouts develop along the plant stem. You harvest from the bottom, up.

■ *Recommended Varieties*

'Long Island Improved' is a taller variety that is less hardy to frost but easier to grow. Earlier to mature and easier to grow, 'Jade Cross' hybrids are dwarf varieties less likely to fall over in the garden. 'Royal Marvel' is another recommended variety.

■ *When and Where to Plant*

Temperature: Brussels sprouts can start growing in the warmer days of summer but will need cool fall weather to mature. In the mountains of North Carolina, plant transplants outside in July. In coastal and

---

GROWING TIP

Brussels sprouts are shallow rooted and prone to falling over. Stake the taller varieties to keep these plants upright and away from the soil.

Piedmont regions, plant transplants or seeds outside in August for growing and maturing for harvest in late winter.

Soil: Add compost to the soil before planting. Brussels sprouts are heavy feeders. Soil pH should be 6.0 to 7.5. Higher pH levels help keep soilborne diseases from affecting the plants.

Sun: Full sun

■ *How to Plant*

Starting seeds indoors: It takes 30 to 45 days to grow plants inside for transplanting outside. Whether you're growing seeds indoors for spring or summer transplanting, you'll need to harden the plants off before planting outside—the plants need to get used to the cold (for spring plantings) and the heat (for summer plantings) before they go in the garden.

Planting outside: Plant transplants outside, 16 to 18 inches apart, or direct-sow into the garden, at 6 inches spacing. Thin seedlings to the 16-inch spacing after plants are 6 inches tall.

■ *How to Grow*

Water: Keep evenly moist throughout the growing season.

Fertilizer: Fertilize with a balanced fertilizer once a month throughout the growing season.

Pest control: Cutworms can affect transplants. To protect young plants, loosely wrap the bottom 3 to 4 inches of stem with newspaper or foil. Cabbageworms can attack larger plant leaves. If you see evidence of cabbageworms, control with *Bacillus thuringiensis* (*B.t.*).

■ *When and How to Harvest*

Sprouts mature from the bottom up. As a sprout begins to form, clip off the leaf below the sprout and let it grow to be ¾ to 1 inch in diameter. Then use clippers to harvest the sprouts as they mature. (You can take enough at a time to eat!)

CABBAGE *(Brassica oleracea)*

Cabbage is a classic cool-season vegetable for Carolina gardeners. In warmer areas, you can grow it from fall through winter and into spring. In the mountains of western North Carolina, cabbage is a fall or spring crop. You can't beat it for utility—it is easy to store and can be added to soups, stir-fries, and cold salads. Along with other cruciferous vegetables, cabbage is one of the most nutritious vegetables you can eat.

■ *Recommended Varieties*

When selecting seeds or transplants, read the plant tag or description and try to select varieties that are resistant to common cabbage diseases such as fusarium wilt, yellows, downy mildew, and black rot. 'Early Jersey Wakefield' is a tasty pointed cabbage. These varieties mature more quickly than round headed types and are sometimes referred to as spring cabbage. 'Savoy Ace' is a savoy leaf variety with wrinkled leaves. 'Ruby Ball', 'Red Acre', and 'Red Rookie' are three red varieties that do well in this area.

■ *When and Where to Plant*

Temperature: Cabbage needs soil temperatures of at least 50 degrees F to germinate and air temperatures of 60 to 65 degrees F. Plant transplants in July in the mountains and August to September for coastal and Piedmont regions. If the temperature yo-yos up and down a lot, the plants will bolt, particularly in spring. In spring, plant transplants outside in March to harvest in late May.

Soil: Cabbage grows best in well-drained, fertile soil. Add compost before planting. The soil pH needs to be at least 6.0 for cabbage to grow. A pH of 6.5 to 7.2 is even better. If the pH is low, add lime to the soil before planting. Higher-pH soils keep club-root disease at bay.

Sun: Full sun

GROWING TIPS

Rotate the cabbage planting areas in your garden, never replanting in the same place more frequently than every two years. Cabbage plants and other cruciferous vegetables are highly susceptible to diseases that linger in the soil. If you have limited space and plan to grow just a few heads per year, consider making a "cabbage barrel" in which you can replace the soil each time you plant new cabbages.

Don't grow butterfly bushes near your cabbage plants. Butterfly bushes can attract pest moths that prey on cabbage plants.

Cabbage is easy to grow and easy to store.

How to Plant

Starting seeds indoors: Sow seeds inside four to six weeks before you plan to plant outside. Harden off transplants before planting in spring.

Planting outside: Because smaller cabbage heads taste better and store more easily, plant cabbage transplants 12 to 18 inches apart in rows 3 feet apart or on 24-inch centers if growing in raised beds. The plants will grow to be 12 to 24 inches in diameter.

How to Grow

Water: Cabbages need evenly moist soil throughout their growing period. Once the heads start to firm up, you can keep the soil on the drier side.

Fertilizer: Sidedress with organic, slow-release fertilizer at the time of planting. One month after planting, apply another application.

Pest control: Look for caterpillars eating the leaves. (They're most likely to show up in fall and spring—not winter.) If you see them, treat with *Bacillus thuringiensis* (*B.t.*) to control them.

When and How to Harvest

The part of the cabbage that you harvest is called a head, and smaller heads taste better. Once the head feels firm, it is ready to pick. Use pruners or a knife to cut off the head just below the base. You can "store" cabbages in the garden for a while, but the longer you leave them in the ground, the more you run the risk of the heads becoming fibrous and bitter.

CARROT *(Daucus carota)*

Once you've eaten a carrot you grew yourself, you will never want to go back to the bland, bitter, soapy-tasting carrots from the grocery store. Carrots actually have flavor, but you'd never know it if you didn't pull one out of your own garden and eat it. These vegetables get a bad rap for being hard to grow. They actually aren't, as long as you prepare the soil well and grow them at the right time of the year—through the winter in the Carolinas. If your garden is sun challenged, you can still grow carrots. They'll produce roots with as little as four hours of sun a day.

◼ *Recommended Varieties*

For carrots that look like typical carrots, grow 'Danvers Half Long', 'Spartan Bonus', 'Scarlet Nantes', 'Apache', or 'Chantenay'. 'Orlando Gold' is a yellow variety. 'Thumbelina' carrots grow fat and round instead of long and skinny. 'Little Finger' carrots are true "baby" carrots, growing to 4 inches or less.

For best results, plant carrots in soil that is loose and free of rocks.

◼ *When and Where to Plant*

Temperature: Carrots will germinate when the soil temperature is 45 to 85 degrees F. They'll grow best at temperatures of 60 to 65 degrees F. They can stay in the garden all winter in warmer regions and all summer in cooler regions. In coastal and Piedmont regions, sow a row of carrots outside every three weeks from July 15 to September 15 and again from January 1 to February 15. In the mountains, sow carrots April 1 to May 1 for a summer harvest and again in July for a fall harvest.

Soil: The most important part of growing carrots is the soil. For carrots to grow long, even roots, the soil has to be loose and free of rocks. If you can grow carrots in raised beds, you'll have fewer problems with the soil being too hard or full of things that get in the roots' way. Turn the soil over and work finely sifted compost into it before planting.

Sun: Full sun to partial shade

GROWING TIPS

Mix carrot seeds with fine sand to make sowing the seeds easier. Because carrot seeds are so small, it's possible to sow too many in one area and waste seeds. Sowing too many also makes for tedious thinning.

Cover the orange tops of the carrot roots if they start to peek out. Exposure to light will make them turn green or brown.

Don't throw out the little carrots that you pull up when you thin the rows—those are real "baby" carrots! Use them in salads.

Too much nitrogen fertilizer will make carrots grow large tops and small roots.

Carrots don't compete well with weeds, so make sure to pull weeds from around carrots as soon as you see them.

◼ *How to Plant*

Starting seeds indoors: Not recommended

Planting outside: Cover seeds with seed-starting mix, and keep the seeds moist while germinating. Carrots can take up to three weeks to germinate, and the soil must stay evenly moist the whole time you're waiting.

◼ *How to Grow*

Water: Keep moist while germinating and evenly watered during the growing season.

Thinning: Thin carrots to 1 inch apart after they sprout and the first true leaves (which look like the second set of leaves) appear. When plants are 3 inches tall, thin to 3-inch spacing.

Fertilizer: Add compost to the garden before planting.

Pest control: Carrots are fairly pest-free in the home garden. The larvae of the swallowtail butterfly will eat the tops sometimes, but this rarely causes enough damage to worry about. Root knot nematodes can be a problem in sandy soils.

◼ *When and How to Harvest*

Carrots take 60 to 75 days to grow to maturity. You can start to harvest them as soon as the roots turn orange (or red or yellow, depending on the variety). Don't let any carrots grow to be larger than 1½ inches wide—they'll start to taste like turpentine at that point. To harvest, just grasp the leafy top and pull. You don't have to harvest every carrot at the same time. Work your way along the row, harvesting what you want to eat at the time of harvest. You can leave carrots in the ground for about a month after they reach full size.

CAULIFLOWER *(Brassica oleracea,* Botrytis Group)

To get white heads on your cauliflower, tie the top of the leaves together over the first small curds.

It's fun to grow your own cauliflower in the garden. It always seems like one day you see nothing but leaves, and you come back the next day to see a cauliflower. It probably isn't that dramatic, but because it takes cauliflowers so long to mature, once you plant them, you tend to ignore them, to some extent. The mature head is somewhat of a surprise. When you think "cauliflower," you probably think of white vegetables, but colored varieties—orange and purple— are available and easier to grow. In the Carolinas, it's best to grow this vegetable in fall.

◼ Recommended Varieties

'Amazing' takes 68 days to maturity and is both heat and cold tolerant. 'Snow Crown' is a fast grower (50 days to maturity) that also tolerates swings in the weather. 'Early Snowball A' is an older self-blanching (does not need to be covered) variety. 'Violet Queen' is a purple variety.

◼ When and Where to Plant

Temperature: Cauliflowers are more sensitive to day/night temperature swings than other plants. In fact, they need cooler nights than days. Seeds will germinate with soil temperatures of 45 to 85 degrees F. Daytime temperatures should be 60 to 70 degrees F and nighttime temperatures 50 to 60 degrees F. You can sow seeds for fall cauliflower crops directly into the garden in July in the mountains or in August to September in coastal and Piedmont regions. Plant transplants outside in August in the mountains and in September in coastal and Piedmont regions.

Soil: Soil pH is important for cauliflower. The ideal pH is 6.0 to 6.5. Cauliflowers are also heavy feeders, so add compost to the garden before planting.

Sun: Full sun

■ *How to Plant*

Starting seeds indoors: It takes a little over a month to grow transplants large enough to plant out in the garden. It's easier to start these plants from seed right in the garden.

Planting outside: Sow seeds 4 inches apart and thin to 12 inches apart after the plants have three sets of leaves. Space transplants 12 to 18 inches apart and plant them 2 to 3 inches deeper than the soil line. The stems will elongate as the plants grow.

■ *How to Grow*

Water: Keep plants evenly moist throughout the growing season. If plants become stressed because of drought or dry soil, they will button and you'll lose the heads.

Fertilizer: Cauliflower is a heavy feeder—particularly of nitrogen and potassium. Sidedress with a balanced fertilizer once a month during the growing season, or apply a slow-release fertilizer at planting time.

Pest control: Cauliflower plants are susceptible to the same pests and diseases that afflict all cabbage-family plants. Protect young plants from cutworms by wrapping their stems with newspaper. Cabbage loopers and cabbageworms attack cauliflower—though less in fall. Keep an eye out for worms. If you see them, you can use *Bacillus thuringiensis* (B.t.) as an organic remedy.

■ *When and How to Harvest*

Check the seed packet or plant tag to determine when your cauliflower plants will be ready for harvest (the number of days from planting to maturity). Once they're ready, pull up the entire plant. Cauliflower heads cannot sit in the garden once they've reached maturity—they'll start to have problems with rot and fungus, and they may bolt.

GROWING TIPS

To get pretty white cauliflower heads, you have to blanch them. Blanching prevents light from reaching the heads, which prevents them from coloring up, hence the term *blanching*, which also means whitening. When the small cauliflower curds begin to form (they will look like baby cauliflowers), use twine to tie the top leaves of the plants together.

Cauliflower is most successfully grown in the fall in coastal areas, when temperatures are more consistent.

COLLARDS *(Brassica oleracea,* Acephala Group*)*

You can't claim to be a Carolina gardener unless you grow collards during winter. This is one of those vegetables that you're more likely to eat if you grow your own. For some reason, fresh-picked collards from the garden seem less stringy and bitter than those you buy at the store. Collards are easy to grow. They're frost tolerant, require no special care (beyond routine pest scouting), and are prolific. Tradition calls for cooking collards in a slow cooker with a ham hock. That's a good way to render their nutritive value null. Try braising collard leaves with garlic, onions, and lemon juice. That's a tastier and healthier way to eat them.

■ Recommended Varieties

The best varieties for the Carolinas include 'Vates', 'Morris's Improved Heading', 'Carolina', 'Georgia', and 'Blue Max'.

■ When and Where to Plant

Temperature: Seeds will germinate in temperatures of 50 to 85 degrees F. Direct-sow outside for a fall or winter crop. Plant seeds in early July in the mountains, mid-July in the Piedmont, and early August in the coastal regions. Start with transplants for a spring crop.

Soil: For the biggest plants, add generous amounts of compost or composted manure to the garden before planting collards. Test the pH to make sure it is 6.0 to 6.5. Add lime if necessary to raise the pH.

Sun: Full sun to partial shade

GROWING TIPS

As a leafy green, collards are heavy feeders and need a steady supply of nitrogen to keep growing. Sidedress with a high-nitrogen fertilizer every four weeks during the growing season.

Rotate the planting area for collards and other cabbage-family plants, and try not to plant in the same place more frequently than every three years.

If you direct-sow collards into the garden, you'll have to thin some of the seedlings to make room for other plants to mature. Don't throw out these young plants! Add them to salads or stir-fries. Smaller collard leaves aren't stringy.

■ *How to Plant*

Starting seeds indoors: It takes six weeks to produce plants large enough to transplant. Decide when you want to plant outside, and start seeds inside accordingly.

Planting outside: Sow seeds outside for a fall or winter harvest. Sow seeds 2 inches apart. When plants are 4 to 6 inches tall, thin plants to 12 inches apart.

■ *How to Grow*

Water: Keep plants evenly moist, never allowing the soil to completely dry out.

Fertilizer: Fertilize every four weeks during the growing season but not the dead of winter, or apply a slow-release fertilizer at planting time and again in February.

Pest control: Use floating row covers to cover collard plants when you see white butterflies (cabbage whites) around the plants. If you see cabbageworms, you can use *Bacillus thuringiensis* (B.t.) to control them.

■ *When and How to Harvest*

Harvest individual leaves and allow the plant to keep growing throughout the season. Start by cutting the outermost leaves first, which will keep the plant growing. Once the plant has matured (in 60 to 70 days), you can also cut the entire plant and bring it inside to process. Collards will keep in the refrigerator crisper drawer for a few weeks, but they taste better and have more nutrients when they're eaten soon after harvesting.

Opposite: Collards are the quintessential Carolina crop.

Once harvested, garlic can be stored in a cool, dry location.

GARLIC *(Allium sativum)*

Garlic, and lots of it, makes everything taste better! It's a misconception that you can't grow garlic in the South. Selecting the right variety makes all the difference, as does planting time. Garlic needs at least 40 days of weather under 40 degrees F for the clove that you plant to split into a bulb with multiple cloves. For this reason, garlic is a fall- and winter-grown vegetable in the Carolinas. If you can purchase garlic bulbs locally, all the better— you're more likely to have success with your plantings.

■ *Recommended Varieties*

There are two types of garlic: softneck and hardneck. Softneck varieties grow best in the hot, humid climate of the South. They do not produce flower stalks, but they are good for braiding, and they store well. Elephant garlic, while not a true garlic, is cultivated the same way that you would cultivate garlic, and it grows well in the South. Hardneck garlic varieties produce flower stalks, or scapes, that are good for pickling, but these varieties are more difficult to grow in the Carolinas.

Good cultivars to plant are the soft-neck Creole varieties 'Burgundy' and 'Ajo Rojo'. Silverskin and artichoke types also grow well in the Carolinas.

■ *When and Where to Plant*

Temperature: Garlic grows best in cool weather, at temperatures of 50 to 60 degrees F. Plant garlic bulbs outside in fall. The bulbs will grow throughout the winter and will be ready to harvest in spring.

Soil: The soil pH needs to be 6.0 to 6.8. Add compost to the soil prior to planting. Garlic grows especially well in raised beds. The extra drainage prevents the bulbs from rotting.

Sun: Full sun

■ *How to Plant*

Starting seeds indoors: Not recommended

Planting outside: Plant garlic cloves 1 inch deep and 4 inches apart.

■ *How to Grow*

Water: Keep the soil moist during the first four weeks after planting—this is when the garlic cloves are growing roots. After that, they do not need extra water until spring. Resume watering once you see leaves start to appear. At the end of the life cycle, decrease watering to allow the bulbs to dry out.

Fertilizer: If you added compost to the soil before planting, you do not need to fertilize while the plants are growing. If you didn't add compost, topdress the area where garlic is planted with an organic fertilizer in the spring.

Pest control: Garlic plants are fairly pest tolerant. In fact, garlic cloves are used in many organic pest control products!

■ *When and How to Harvest*

Garlic is ready to harvest when the tops start dying back. When you see the leaves start to turn yellow, stop watering the bulbs. Once the tops have died back, use a trowel or soil knife to gently dig up the bulbs. (Sometimes you can pull the bulbs up by the leaves, but not always.) You can store garlic by braiding the tops together and hanging in a cool, dry location.

GROWING TIPS

Don't try to plant garlic that you purchase at the grocery store. It might not be the right variety for your area, and it also might have been treated to prevent sprouting.

Keep the garlic patch weeded—garlic bulbs don't compete well with weeds.

KALE *(Brassica oleracea,* Acephala Group*)*

Kale has become an "it" vegetable. Everyone is drinking kale smoothies, eating kale chips, and generally espousing the benefits of this leafy green vegetable. It's for good reason, though: kale is the most nutrient-dense vegetable per calorie you can eat. It's also easy to grow. Once you plant it, you can basically leave it alone and harvest from the plants for months. Kale even grows through the summer in the Carolinas, though the flavor is not the same as winter-grown kale. Young leaves are tender enough to put in fresh salads, while larger leaves are perfect for soups and braised greens.

For a tasty and healthy treat, prepare a large bunch of kale by removing the center ribs and tearing the leaves into 1- to 2-inch pieces. Place in a bowl with a sprinkle of salt, a bit of olive oil, and a generous soaking of flavored vinegar. You'll find yourself craving this delicious salad and searching for new flavored vinegars to try it with.

■ *Recommended Varieties*

Recommended varieties include 'Green Curled Scotch', 'Early Siberian', 'Vates, Dwarf Blue Curled Scotch', and 'Blue Knight'. 'Redbor' is a purple variety with large, curly, purple leaves and veins. 'Red Russian' has green leaves with a red vein. The young leaves of 'Red Russian' are great in salads. 'Toscano' is a dinosaur-type kale with long, strappy, wrinkly leaves. It holds up well in cooking. Kale is kale, though. Get your hands on some seeds and transplants, plant at the right time, and you're almost guaranteed success.

■ *When and Where to Plant*

Temperature: Kale seeds will germinate in temperatures of 45 to 95 degrees F. Kale grows best with daytime temperatures of 60 to 65 degrees F. However, kale is tough and will withstand both higher temperatures if it has enough water and low temperatures. Sow seeds directly outside in July in the mountains and in August or September in coastal and Piedmont regions for a fall or winter harvest. Replant transplants outside in February or March in coastal regions, in April in the Piedmont, and in May in the mountains for a spring harvest.

Soil: Add compost to the soil before planting. Test the soil to make sure that the pH is 6.0 to 7.0.

Sun: Full sun to partial shade; will last longer into summer in partial shade

■ *How to Plant*

Starting seeds indoors: Start seeds indoors at least one month before you want to transplant outdoors.

Planting outside: Plant transplants outside with at least 6 inches between plants. Sow seeds with two inches between plants and thin as the plants grow.

■ *How to Grow*

Water: Keep kale growing, thriving, and producing tender (not bitter and fibrous) leaves by watering consistently when the plants are actively growing.

Fertilizer: As with all green, leafy vegetables, kale is a high-nitrogen feeder. Fertilize by applying an organic, slow-release fertilizer at planting time and again in February (for fall crops) or midsummer for summer crops.

Pest control: Kale is much hardier and much more resistant to pests than most other plants in the cabbage family. Aphids can be a problem. Control these with insecticidal soap. Cabbageworms and cabbage loopers can also munch on kale. If you see the worms, treat with *Bacillus thuringiensis* (*B.t.*).

■ *When and How to Harvest*

Keep your scissors handy when harvesting kale. A sharp yank can cause the whole plant to come out of the ground. Cut small young leaves (1 inch long or less) to use in fresh salads. Harvest the larger leaves from the bottom of the plant upward. This lets inner leaves keep growing. You'll know when it's time to stop harvesting by the taste of the leaves. Once they get to be too stringy or woody, it's time to dig the plants up.

GROWING TIPS

Kale tastes better when harvested in spring or fall. You can keep plants growing through summer, but wait for cooler weather to arrive before resuming harvesting.

Kale is easy to grow and a popular edible.

KOHLRABI (*Brassica oleracea,* Gongylodes Group)

The stem of the funny-looking cabbage-family vegetable kohlrabi is the part you eat. It swells up like a bulb and is crunchy and juicy. It's originally from eastern Europe and tastes like a mild turnip, but it isn't a root vegetable. You can cook kohlrabi as you'd cook potatoes or turnips, though it's also good raw—you can use grated kohlrabi in salads or slaws. As part of the cabbage family, kohlrabi is another highly nutritious vegetable that's easy to grow in spring and fall in the Carolinas.

The swollen stem of the kohlrabi plant is what we eat.

■ *Recommended Varieties*

'White Vienna' and 'Grand Duke Hybrid' are two white varieties. 'Blaro' and 'Early Purple Vienna' are purple varieties.

■ *When and Where to Plant*

Temperature: Kohlrabi needs a minimum soil temperature of 55 degrees F to germinate. It is the least frost hardy of all cabbage-family plants, so plan to plant it so that it can mature before nighttime temperatures are routinely in the 40s F. For spring harvests, sow seeds or set transplants outside from March 1 to April 15 in coastal and Piedmont regions, and from April 15 to May 15 in the mountains. For fall harvests, sow seeds in the garden from August 1 to September 15 in coastal and Piedmont regions, and from July 1 to August 1 in the mountains.

Soil: Plants thrive in soil with a pH of 6.0 to 6.5. Add compost to the soil before planting.

Sun: Full sun, especially during cooler weather

■ *How to Plant*

Starting seeds indoors: You can start seeds indoors a month before you want to plant outside. However, it is just as easy to grow kohlrabi from seeds sown directly into the garden.

Planting outside: Plant kohlrabi seeds every three weeks during the planting season to enjoy successive harvests. Most varieties mature within 40 to 50 days, so planting frequently will allow you to harvest a fresh crop for several months.

■ *How to Grow*

Water: Keep seeds and seedlings moist while germinating; then monitor and keep evenly moist throughout the growing season.

Fertilizer: Kohlrabi is not a heavy feeder but will benefit from sidedressing with a balanced fertilizer midway through the growing season.

Pest control: Kohlrabi is more resistant to pests than some cabbage family members, but it still has problems sometimes with flea beetles and cabbageworms. You can use Neem oil to control flea beetles. Floating row covers can prevent flea beetle damage and damage from cabbageworms. If you see worms, spray with *Bacillus thuringiensis* (*B.t.*).

■ *When and How to Harvest*

Harvest kohlrabi by pulling up the entire plant. Start harvesting when the stems are 1 inch in diameter, and continue to harvest until the stems are 3 inches in diameter. If you let the stems grow any larger, the vegetables will be woody, fibrous, and tough. This is a vegetable that can confuse gardeners, because the vegetables you buy at the grocery store are sometimes much larger than what you'll harvest at home. Don't be tempted to let your kohlrabi grow too long to match your idea of what it's supposed to look like.

GROWING TIPS

If you are sowing seeds in summer for a fall harvest, provide the seedlings with some shade to protect them from the scorching sun.

Use a blanket or sheet to protect plants from cold weather, including late-spring and early-fall frosts. Temperatures in the 40s F after the plants have been growing a while can cause the plants to bolt and flower, resulting in a loss of the crop.

LEEK *(Allium ampeloprasum,* Porrum Group*)*

I love leeks. They're so easy to grow, and they're fun to cook with. You can throw young, tender leeks on the grill. Or you can cut up larger leeks to make potato and leek soup. (Sauté 8 cups of chopped-up leeks until soft. Add 8 cups of diced potatoes and cover with chicken broth. Simmer until the potatoes are tender. Purée the vegetables. Pour in 1 cup of heavy cream. Delicious!) One thing to be careful about when cooking with leeks: because you blanch the stems by hilling the leeks (see Growing Tip), a lot of soil and grit can get stuck between the leaves. I always cut them in half lengthwise and spend quality time washing them before chopping them up. Then I wash them again after chopping, tossing them in a colander to get any remaining grit out.

◼ *Recommended Varieties*

'Lancelot' is a good variety to plant in fall and grow through the winter. It is one of the most cold-tolerant types. 'Arkansas' and 'Tadorna' will also overwinter in our area.

◼ *When and Where to Plant*

Temperature: Leeks take a long time to mature, and they require cool temperatures to grow. It's best to plant leeks in fall and let them grow throughout the winter. In coastal and Piedmont regions, sow seeds outside from August 1 to September 15 for harvesting from January 1 to March 1. You can also plant transplants outside from September 15 to October 31 and again in February. In the mountains, plant transplants outside right after the last frost date.

Soil: Add compost to the soil before planting. Leeks grow best in loose soils full of organic matter. Test the soil to ensure a pH of 6.5 to 7.0.

Sun: Full sun

GROWING TIPS

To get long, white stems, you have to hill leeks. As the plants grow, pile the soil up around the bottom leaves—just to the bottom of the leaf fork. This will encourage the plant to keep growing upward and forming new leaves at the top of the plant.

An alternative method in well-drained soils is to plant leeks in a shallow trench (4 to 6 inches deep). As the leeks grow, fill in the trench.

Keep the leek patch weeded. The one thing leeks are picky about is weeds.

■ *How to Plant*

Starting seeds indoors: Start seeds indoors eight to ten weeks before you plan to plant leeks outdoors. You can sow eight to ten seeds per 4-inch pot and then tease apart the roots when you're ready to plant outside.

Planting outside: Sow seeds 2 inches apart outside and thin to 6 inches after plants have two sets of leaves. When planting transplants, harden off before planting in the garden.

■ *How to Grow*

Water: For the best flavor, keep leeks evenly moist throughout the growing season. Do not let them dry out.

Fertilizer: Leeks are not heavy feeders. Sidedress with a balanced fertilizer every two months during the growing season but not during the dead of winter.

Pest control: Leeks are fairly pest-free.

Both the white root and the green leaves of the leeks can be eaten.

■ *When and How to Harvest*

Leeks mature in 100 to 120 days, depending on the variety. Once the stalks reach ½ inch in diameter, you can start harvesting them, though. If you've planted a lot of leeks (a 4-square-foot patch, for example), you'll have plenty of plants to harvest over a period of time. You can pull the younger plants to use in salads, omelets, and other quick-cook meals and save some plants to grow larger for use in soups and stocks. Harvest leeks by pulling up the entire plant.

Most people cook with the tender, mild white part of the leeks. You can use the green tops for flavoring broth and soups, too.

LETTUCE *(Lactuca sativa)*

Lettuce is the quintessential cool-season crop. It isn't just the cool weather that lettuce likes, though. Lettuce needs the short days (and long nights) associated with fall, winter, and early spring to stay in a vegetative state—producing leaves not flowers. Once the weather heats up, lettuce leaves get tough and bitter, and plants will send up a flower stalk. Pollinators of your summer vegetables love lettuce flowers, so once you're done harvesting for leaves, let some of the plants flower to draw beneficial insects to your garden. Once you start growing your own lettuce, you'll be amazed at the amount of money you save. Buying fresh looseleaf lettuce at the market is expensive!

■ Recommended Varieties

There are four types of lettuce: crisphead (Iceberg types), butterhead, romaine, and looseleaf. Butterhead and looseleaf are the easiest types to grow in the Carolinas. Crisphead doesn't do well in our heat, but romaine lettuce can be grown, especially in the fall. 'Black-Seeded Simpson' and 'Simpson Elite' are two good green-leaf varieties. 'Red Sails' and 'Lolla Rosa' are easy-to-grow red-leaf varieties. Try 'Buttercrunch' if you want a heading lettuce. You can find a variety of mesclun and cut-and-come-again seed mixes at garden centers and home-improvement stores—all of these will do fine if grown in fall or spring in the Carolinas.

■ When and Where to Plant

Temperature: Lettuce is a short-day, cool-season crop. For a fall planting, sow seeds outside from August 15 to September 1 in all regions. Sow outdoors from December 15 to March 15 for winter and spring harvesting in coastal regions. Sow outdoors from February 15 to April 1 in the Piedmont and in April in the mountains.

Soil: Lettuce needs a soil pH of 6.0 to 7.0. Because it is a heavy feeder, you can save yourself some work by adding compost to the soil before planting.

Sun: Full to partial sun; sometimes lettuces will last longer into the summer if planted in partial shade.

Use scissors to snip off lettuce leaves, and it will grow back.

GROWING TIPS

Cover lettuce seeds planted outside with seed-starting mix rather than garden soil. It is hard for tender lettuce seedlings to break through a hard soil crust.

Sow some lettuce seeds around summer plants such as tomatoes and peppers. You can get a second late-spring crop of lettuce this way.

Lettuce is shallow rooted, so be careful when harvesting. It's better to cut leaf lettuce than to pull or twist off leaves. You don't want to yank the whole plant out of the ground.

■ *How to Plant*

Starting seeds indoors: You can start lettuce seeds indoors, but it's so easy to grow lettuce from seed outdoors that it really isn't worth the time or hassle to start indoors.

Planting outside: Sow seeds outdoors and cover with seed-starting mix. Press the soil down and water. (Pressing the soil cover down helps keep the seeds from floating away.) Sow head lettuce seeds with 1 inch between seeds. Thin to 4 inches after the plants have three sets of leaves. Thin to 12-inch spacing as the plants start to mature. Eat what you thin! Sow cut-and-come-again lettuces and leaf lettuces thickly—three seeds per inch.

■ *How to Grow*

Water: Keep the soil evenly moist.

Fertilizer: Lettuce likes to eat. As with all leaf crops, lettuce responds well to applications of composted chicken manure.

Pest control: Slugs are a major problem with lettuce. Spread diatomaceous earth around lettuce plants to create a barrier for slugs.

■ *When and How to Harvest*

Harvest head lettuce by pulling up the entire plant. Harvest leaf lettuce by using scissors to snip off leaves about ½ to 1 inch above the soil line. The leaf lettuce will grow back. Always harvest in the morning when the water content of the leaves is highest. I like to use a salad spinner to wash homegrown lettuce. Fill the bowl with cold water and swish the leaves around. Dump them out into the strainer to drain. Repeat. Then spin the water away from the leaves, drain the bowl, and keep the lettuce in the salad spinner in the refrigerator. The lettuce will remain crunchy and crisp.

ONION *(Allium cepa)*

You almost can't cook without onions. They're so inexpensive to buy that you might wonder why you should bother growing them. The reason to consider growing onions is that, if you select the right varieties, they're fairly easy to grow throughout the winter and require little care. You can grow onions from seeds, bulbs (also called sets), and transplants. There are short-, long-, and intermediate-day onions. In the Carolinas, short-

For best results, grow short-day onions in the Carolinas.

day varieties are best. Bunching onions, also called green onions, never form bulbs and can grow anywhere. The sweet, tasty Vidalia-type onions that are perfect for eating on hamburgers or cooking in soup are short-day onions.

■ *Recommended Varieties*

If you want to harvest bulbing onions for storage in the Carolinas, you need to grow short-day onions during winter. Suggested varieties include Candy, Granex (white), Stockton Sweet Red, and Yellow Granex. If you can't find these varieties, be sure to buy short-day onions. You cannot grow intermediate-day and long-day onions in the Carolinas, because it is too hot during summer. Read the package carefully before purchasing onions to make sure you get the right ones. You can grow bunching onions in the Carolinas in spring, including Beltsville Bunching and Evergreen Bunching.

■ *When and Where to Plant*

Temperature: Onions are cool-season crops that will grow best in fall, winter, and spring. Do not attempt to grow onions in summer.

Soil: Onions grow best in loamy soils. You'll have to fertilize more if you are growing onions in sandy soil. Clay soils produce more pungent onions.

Sun: Full sun

■ *How to Plant*

Starting seeds indoors: Start onion seeds indoors six weeks before you plan to transplant them outdoors.

Planting outside: You can plant seeds outside in coastal and Piedmont regions from September 15 to October 30. Plant transplants outdoors (after hardening off) from February 1 to March 15 in coastal and Piedmont regions. Plant transplants outdoors in late March and early April in mountain areas. Plant sets outside in October and February in coastal regions, September and March in the Piedmont, and August and April in the mountains.

How to Grow

Water: Onions don't require a lot of water, but they do need even moisture. Dry soil will cause the onion to produce two bulbs instead of one. Increase watering as the bulb begins to form and swell. Decrease watering in the last two weeks before harvesting.

Fertilizer: Incorporate a balanced, slow-release fertilizer into the planting bed before planting onions. If you are growing onions in sandy soil, sidedress every four weeks during the growing season only, not in the dead of winter. High-nitrogen fertilizer can cause onions to be less winter hardy and produce smaller bulbs.

Pest control: Onion root maggots are the most problematic pests for onions. Good sanitation and removal of dead plant leaves will help control onion maggots.

When and How to Harvest

Harvest bunching onions when they are ⅓ inch in diameter. Pull up the entire plant, and chop it up for use in salads, soups, stews, and sandwiches.

Bulbing onions have a more delicate harvesting process. A month before you want to harvest the onions, start gradually pulling the soil away from the bulbs. By harvest time, the bulbs should be one-third uncovered. Pull up the onions when one-third of the tops have fallen over. Lay them on top of the soil and allow them to cure (dry out) for three to five days. Cover the plants if rain is forecast. Before bringing the onions inside, cut the tops off, leaving 1 inch of stalk to dry on the plants. You can store short-day onions for two to three months in a cool environment with low humidity.

GROWING TIPS

Make sure you are purchasing short-day onions. You won't be able to grow long-day onions to maturity in the Carolinas. Mulch onions to keep weeds down, soil temperature cool, and moisture levels even. Eat immature bulbing onions that you pull up during thinning as you would eat green onions. Don't add sulfur to the soil where you're growing onions—it will cause the onions to have an off taste.

PARSNIP *(Pastinaca sativa)*

Parsnips are relatives of carrots, but they're sweeter with a buttery, smooth flavor. You don't want to eat parsnips raw, but they taste great cooked, either roasted or in soups and stews. These cold-hardy vegetables require many of the same growing conditions as carrots. They're high in vitamin C, fiber, and folate, and they're easy to grow once they germinate. Put these in your "plant and forget" category of vegetables.

■ *Recommended Varieties*

There aren't a lot of varieties of parsnips, but 'Javelin' is one that does well and is cold tolerant for overwintering in the Carolinas.

■ *When and Where to Plant*

Temperature: Parsnips grow best during fall/winter or spring in the Carolinas. Sow seeds directly outside in July in the mountains and in August to September in coastal and Piedmont regions. Sow seeds outside again in mid-February to mid-March in coastal and Piedmont regions and in April in the mountains.

Soil: Parsnips need a pH of 6.0 to 6.5 and deep, workable, loose soil. Hard clay soils can cause forked roots. Dig compost into the planting bed before planting.

Sun: Plant in full sun to partial shade

Like carrots, parsnips need deep, loose soil to grow well.

GROWING TIPS

Buy seeds packed for your growing year. Parsnip seeds have a short shelf life.

Parsnips can take a while to germinate. Make sure the bed where you planted the seeds stays consistently moist while you're waiting for the plants to sprout.

Parsnip leaves can cause a skin rash in some people. Always wear long sleeves and gloves when handling the plants.

◼ *How to Plant*

Starting seeds indoors: Not recommended

Planting outside: Soak seeds for twenty-four hours before planting. Sow seeds thickly outside—at a rate of three or four seeds per inch. Cover with seed-starting mix, and water. You might want to cover with frost cloth to help keep the soil from drying out during germination. Parsnips take up to 20 days to germinate, and the soil can't completely dry out during that time. Thin seedlings to 4 inches apart once they germinate and have two sets of leaves.

◼ *How to Grow*

Water: Keep the soil evenly moist, and water deeply so that the top 6 inches of soil are soaked when you water.

Fertilizer: Sidedress with a balanced fertilizer once a month during the growing season, or apply a slow-release fertilizer at planting time.

Pest control: Parsnips are fairly pest-free. If rabbits and squirrels are a problem in your area, use a floating row cover for young plants. Swallowtail caterpillars will sometimes feast on the plant leaves. You can pick these off by hand.

◼ *When and How to Harvest*

Parsnips are ready to harvest when the root tops are 1½ to 2 inches in diameter. Use a garden fork to loosen the soil around the parsnips and help dig them up. It takes 120 to 150 days for parsnips to grow from seed to maturity, and by that time they're well rooted in. Parsnips always taste better when they've been exposed to cool weather.

PEA *(Pisum sativum)*

Everyone can grow peas. They are, hands down, the easiest vegetable to grow. You can grow shelling peas (that you remove from the shell before eating) and edible-podded peas (such as snap or snow peas). Pea shoots, the top 6 inches of growth on the pea plant, are also edible. Cook pea shoots like you'd cook braising greens, or sauté them in olive oil and lemon juice with a bit of garlic and toss with pasta. Yum! Peas are another vegetable that will save you a lot of money if you grow your own. It's expensive to buy tasty fresh peas but inexpensive to grow them.

Growing your own peas can save you money at the grocery store.

■ *Recommended Varieties*

If you want to pick delicious peas and eat them right out of the garden, grow sugar snap or snow peas. Recommended varieties are 'Sugar Snap', 'Mammoth Melting Sugar', 'Snowbird', and 'Sugar Bon'. Try these varieties of shelling peas: 'Wando', 'Green Arrow', 'Freezonian', and 'Tall Telephone'.

■ *When and Where to Plant*

Temperature: Plant peas in early spring in all areas. If you plant two successive groups of peas, they'll ripen at different times so that you're not overwhelmed with more peas than you know what to do with. In the mountains, plant peas in March or early April as soon as the soil is workable. In coastal and Piedmont areas, plant peas from February 1 to March 1.

Soil: Peas do not need excellent soil. As legumes, they fix their own nitrogen—in effect making their own food.

Sun: Full sun

■ *How to Plant*

Starting seeds indoors: Not recommended

Planting outside: Use a hoe to dig rows 1 inch deep (even in raised beds). Plant rows 8 inches apart in raised beds and 18 inches apart in regular garden beds. Space seeds 2 inches apart when planting.

■ *How to Grow*

Water: Keep the soil evenly moist, but not soaking wet. Do not allow the bed to completely dry out.

Fertilizer: Do not over-fertilize peas with nitrogen. Too much nitrogen can cause lots of green growth and no flowers. No flowers equals no peas! If you have poor soil, sidedress with a lower-nitrogen fertilizer when plants are 6 inches tall. The phosphorus and potassium will encourage good fruit development, and that's what you want to eat—the fruits!

Pest control: Aphids are the biggest pest problem for peas. Use insecticidal soap according to package instructions to control the pests.

■ *When and How to Harvest*

Pick or clip off individual pea pods to harvest. Peas will be ready for harvest about 60 days after planting.

Harvest sugar snap peas when the pods are plump but still dark green in color. Harvest snow peas when the pods are still relatively flat. The flowers might still be hanging on to the ends of the pods of both types of peas when you harvest the pods.

Harvest shelling peas when the pods are plump but still dark green. Once the pods start to turn yellow, the sugars in the peas convert to starch, and they aren't tasty anymore.

GROWING TIPS

Place pea stakes or supports as soon as you see the plants germinating. They grow fast and it's nearly impossible to wrestle a staking system into place once the plants are taller than 6 inches.

If you have poor germination rates, the peas could have succumbed to a fungal problem. Try again with a different variety that's disease resistant, or plant in a different area.

If you have tall plants but few peas, pinch off the growing tip of the plant to encourage more fruit production.

RADISH *(Raphanus sativus)*

Radishes are cool-season root vegetables. The roots are the most commonly eaten part of the radish, but the tops are edible, too. Spring radishes are fast to germinate and mature. You can get several radish crops from one package of seeds. Radishes are beneficial plants for the vegetable garden, too, because they deter pests and, when allowed to flower, attract beneficial insects and pollinators. It's a good idea to sow successive crops of radishes so that they mature a few weeks apart. Nobody needs sixty radishes at once!

■ Recommended Varieties

You can find infinite varieties of radishes in the store. 'Cherry Belle' is the typical round, red-on-the-outside, white-on-the-inside variety that most people think of when they think "radish." 'French Breakfast' radishes have longer roots. You can also purchase seed mixes with red, white, purple, and yellow radishes in them.

Because radishes grow so quickly, they are a good plant for children to grow.

GROWING TIPS

Plant successive crops of radishes so that they don't mature all at once.

Let some radishes stay in the garden to flower. This will attract beneficial insects and pollinators to your summer crops.

■ *When and Where to Plant*

Temperature: The soil temperature has to be at least 50 degrees F for radishes to germinate. It is tempting to plant spring radishes too early, but if the soil is still cold, they won't sprout.

Soil: Radishes are not picky about soil, but, as root crops, they will rot if the soil is slow draining or wet.

Sun: Full sun

■ *How to Plant*

Starting seeds indoors: Not recommended

Planting outside: Sow successive plantings of radish seeds outside as soon as the soil can be worked and has drained. Plant seeds ½ inch deep and 2 inches apart. Radishes have a high germination rate. The more space you put between the seeds, the less thinning you'll have to do.

■ *How to Grow*

Water: Radishes aren't fussy, but they do need even moisture levels to avoid cracking.

Fertilizer: To grow larger roots, you can fertilize with fish emulsion two weeks after planting.

Pest control: Use row covers to prevent flea beetles and aphids from attacking the plants. You can also spray with insecticidal soap for aphids.

■ *When and How to Harvest*

Don't let the radishes linger. They taste best when harvested after three to four weeks of growth, when they are still relatively small. The older the radishes, the more fibrous and stringy they get, and the more pungent the flavor becomes.

SPINACH *(Spinaceia oleracea)*

Why should you grow your own spinach? Because field-grown and harvested spinach is easily contaminated, and, if you like to eat your spinach raw, that can cause unpleasant intestinal problems. Spinach is a cool-season green grown for its leaves. You can actually overwinter spinach, even in the mountains, if it is planted with enough time to grow to some size before winter. Spinach is a long-day flowering plant, which means it stays vegetative during the short days of fall, winter, and spring. As soon as the temperature starts to warm up and the days lengthen, spinach will start to bolt, or flower, and the season's over.

■ *Recommended Varieties*

There are different types of spinach. Savoy types have dark green, crinkled leaves. There are also flat-leaf types, and semi-savoy types that have somewhat wrinkly leaves. It doesn't matter which variety you grow. Recommended varieties of the three types include 'Hybrid 7' (semi-savoy), 'Dark Green Bloomsdale' (semi-savoy), 'Tyee Hybrid' (semi-savoy), 'Melody' (savoy), and 'Space' (semi-savoy).

■ *When and Where to Plant*

Temperature: Spinach will germinate in soil temperatures of 45 to 75 degrees F. It grows best with air temperatures of 60 to 65 degrees F. Plant seeds outside in coastal and Piedmont areas from September 15 to October 15 in two-week successions. Sow outside in August in the mountains. Plant transplants outside after hardening off in coastal and Piedmont areas from mid-February to early April.

GROWING TIPS

Sow successive plantings to ensure fresh harvests of tender young leaves over a longer period of time.

Use drip irrigation or soaker hoses to provide steady water at the root zone and away from the leaves. Fungal diseases that plague spinach can spread through water on the leaves.

Once spinach sends up a flower stalk, pull up the plant, and compost it. There's no use in trying to keep the plant in a vegetative state once it has moved on.

Spinach is hardy but can be affected by hard frosts. If temperatures are forecast to be below 28 degrees F, cover with frost cloth.

Soil: Spinach needs a soil with a pH of 6.0 to 7.5. It is a leaf crop and as such is a heavy feeder. Add compost to the soil before planting to ensure healthy growth.

Sun: Full sun

How to Plant

Starting seeds indoors: Start seeds indoors a month before you want to transplant outdoors. Working with transplants is easier for spring crops than for fall crops.

Planting outside: Soak seeds overnight before planting outside. Sow fall crops directly outside into the garden to avoid problems with bolting. Space seeds 2 inches apart and thin to 6 inches apart after plants have three sets of leaves. Harden off transplants and plant 6 inches apart.

With a little planning, you can overwinter spinach anywhere in the Carolinas.

How to Grow

Water: Spinach is not a heavy drinker, but keep the soil evenly moist, and avoid wet/dry/wet/dry situations.

Fertilizer: Spinach is a leaf crop and a heavy feeder. Sidedress with a balanced fertilizer every two weeks during the growing season.

Pest control: Prevent fungal problems by keeping water off the leaves. Use floating row covers to prevent damage from leaf miners.

When and How to Harvest

You can start harvesting spinach leaves three weeks after planting or when the plants have at least ten leaves on them. Cut the outermost leaves first, always leaving six leaves on the plant to produce sugars to help the plant keep growing. Use scissors or pinch off the leaves at the base of the leaf stalk. Spinach matures in about 40 days. You can also wait until that point and harvest the entire plant.

SWISS CHARD *(Beta vulgaris chicla)*

Swiss chard is a biennial green related to beets. Both the colorful leaves and the stalks of the plants are edible. It will grow through the summer if planted during cool weather, making it one of those hard-to-define vegetables, in terms of seasons. Swiss chard is loaded with vitamins and is multipurpose in the kitchen. Young leaves can be picked and added raw to salad. Larger leaves can be added to soups and stews, omelets, sandwiches, and stir-fries. Swiss chard is so pretty that it doesn't have to be contained in the vegetable garden. If you're short on space, grow this beautiful vegetable in the perennial flower garden. It will look right at home.

■ *Recommended Varieties*

'Bright Lights' is a multicolored mixture of plants with stems ranging from yellow to white to red to green. 'Rainbow' is another multicolored mix. 'Fordhook Giant' produces large green leaves with white stems.

Swiss chard is pretty enough to grow in a flower garden.

■ *When and Where to Plant*

Temperature: Swiss chard grows best in cool temperatures. Plant outside in early April to mid-May in the mountains. Plant outside in October to November and again in March to April in coastal and Piedmont regions.

Soil: Swiss chard is not a picky plant and will grow in almost any soil. Add compost before planting, if possible.

Sun: Full sun to partial shade

■ *How to Plant*

Starting seeds indoors: Not necessary

Planting outside: If you plant transplants purchased from the garden center, space plants at least 8 inches apart. Swiss chard grows to be fairly large. To sow seeds directly out into the garden, soak seeds overnight before planting. Plant one seed every 2 inches, ½ inch deep.

■ *How to Grow*

Water: Keep chard evenly moist.

Fertilizer: Fertilize with a balanced fertilizer every two to four weeks during the growing season.

Pest control: Use insecticidal soap to take care of any aphids or leaf miners that might attack.

■ *When and How to Harvest*

You can start harvesting Swiss chard leaves as soon as the plant has at least four or five leaves. Take the outermost leaves first, and allow the inner leaves to grow and provide sugars for the pant. Use a sharp knife, scissors, or hand pruners to cut the leaf stalks at the soil level.

GROWING TIPS

Thin plants once they are 3 or 4 inches tall so that there's one plant every 6 inches.

Leave Swiss chard in the garden throughout the summer, taking care to give it extra water. This is one green that you can grow virtually year-round in most areas of the Carolinas.

TURNIP *(Brassica rapa* ssp. *rapa)*

Turnip greens are a Southern Sunday-table staple, but we also grow turnips for their roots. They're a cool-weather vegetable that grows well in early spring in the Carolinas. If you time it right, you can get a fall crop out of the garden, too. When you think of a turnip, you probably think of large, purple or white, globe-shaped roots, but there are other smaller, sweeter varieties, too. If you've never thought much about turnips, give them a try. They're naturally high in vitamin C and easy to grow.

■ *Recommended Varieties*

Traditional large turnip varieties include 'Purple Top', 'White Globe', and 'White Lady'. The 'All Tops' variety is grown specifically for turnip greens. 'Hakurei' is a small, white salad turnip that tastes good when shredded raw into salads.

■ *When and Where to Plant*

Temperature: Turnips will sprout when the weather is warm, but they need cool temperatures to mature if you want to harvest nice roots. Sow seeds directly into the garden from February 15 to March 15 in coastal

Turnips are grown both for their leafy green tops and their bulbous roots.

GROWING TIPS

Plant turnips every three weeks during the optimum planting season for a staggered harvest.

Thin turnips to 4-inch spacing when the plants are 4 inches tall. You can eat the young greens that you pull up. Harvest every other plant when the roots are 1 inch in diameter. This will give you fresh turnips to eat immediately while allowing room for half of the harvest to grow larger roots.

regions and again from August 15 to October 15. Sow seeds outside from March 15 to April 15 and again in August in the Piedmont, and from March 15 to May 15 in the mountains.

Soil: The biggest concern with turnips is to avoid planting them in exactly the same place that you grew cabbage and cabbage-family plants.

Sun: Full sun

How to Plant

Starting seeds indoors: Not recommended

Planting outside: Sow turnip seeds thinly (two seeds per inch) outside and cover with seed-starting mix. Turnips germinate quickly and will sprout within a few days.

How to Grow

Water: Keep the soil evenly moist but not soggy.

Fertilizer: Add compost to the bed before planting and you won't have to add additional fertilizer. If you have sandy soil, fertilize once with a balanced fertilizer about three weeks after planting.

Pest control: Aphids can be controlled with insecticidal soap. Control flea beetles by using row covers.

When and How to Harvest

How you harvest depends on which part of the plant you want to harvest. Young, tender greens can be picked for salads when they are 2 to 4 inches tall. Pull up the young, tender roots when they are 2 inches in diameter and use them raw in salads. When the larger roots grow to 3 or 4 inches in diameter, they're ready to be pulled up and used for roasting and mashing. The greens can be braised or added to soups. Turnips have a sweeter flavor when allowed to mature when the weather is cooler.

Cool-season herbs cut for use as garnishes.

Cool-Season Herbs

Many of our favorite herbs grow best when the weather is cool. Dill, fennel, and cilantro are just a few of the everyday herbs we use that thrive during fall, winter, and early spring. Some of these herbs will hang on and grow in summer if provided with protection from the sun while others just physiologically won't stay vegetative once the days get longer—their fragrant leaves give way to flowers and seeds. Don't miss out on your favorite seasonings. Plant the herbs covered in this section during the cool-weather growing season.

Parsley

CILANTRO/CORIANDER
(Coriandrum sativum)

Cilantro is an annual cool-season herb prized for its fragrant leaves. Either you love cilantro or you hate it. I love it. I just wish I could grow big, bountiful bunches at the same time that tomatoes are ripening, but I can't in my coastal North Carolina garden. This parsley-family plant grows best in our area in early spring. Once the days get longer and hotter, it bolts, sending up a flower stalk that produces seeds. Cilantro seeds are the herb that we call coriander. Don't frustrate yourself growing this herb when the weather is warm. Plant it in spring, and learn how to enjoy it with hearty cool-weather fare.

We use the leaves of cilantro in salsas, but they are also tasty in salads and curries.

When and Where to Plant

Temperature: Cilantro grows best in temperatures of 50 to 85 degrees F. Sow seeds outside in February to March in coastal and Piedmont regions. In the mountains, sow outside in early April.

Soil: The herb grows well in soils that are slightly alkaline, with a pH of 6.5 to 7.5.

Sun: Full sun to partial shadeL Planting in partial shade will extend the harvest window slightly.

How to Plant

Starting seeds indoors: Not recommended

Planting outside: Plant seeds ¼ inch deep, with two seeds per inch.

How to Grow

Water: Keep seeds moist while sprouting, and avoid wet/dry/wet/dry periods while growing.

Fertilizer: Fertilize with a balanced fertilizer two weeks after germination.

Pest control: Cilantro has few pest problems but is susceptible to some fungal diseases spread through splashing water. Try to water the plants at the root zone and avoid splashing the water.

When and How to Harvest

To harvest leaves, snip the top off the plant. The smallest, most immature leaves have the best flavor. If you want to save coriander seeds, pull up the whole plant once seeds have formed and the plant has turned yellow. Hang it upside down to dry over newspaper to collect the seeds.

GROWING TIP

Stay on top of weeds while plants are germinating. Because the seeds are slow to germinate, if you let the weeds get out of control, the plants will have a hard time growing. The weeds will outcompete them for water and sun.

Dill leaves can be used in salads, and the seeds in flavoring.

DILL *(Anethum graveolens)*

I love to pick dill at the same time I'm harvesting salad greens. I wash and keep the dill in the same salad spinner as the lettuce. Then I have herb-flavored salad mix for none of the cost of buying it at the grocery store. Dill is an important addition to the vegetable garden, as it serves as a host plant for native black swallowtail butterfly caterpillars, and a its flowers are a nectar source for beneficial insects and pollinators.

When and Where to Plant

Temperature: Seeds will germinate with soil temperatures of 50 to 70 degrees F. While the plant likes to grow during cool weather, it is frost sensitive. Plant dill seeds immediately after danger of frost has passed.

Soil: Like its relative the carrot, dill needs well-drained, loose soil.

Sun: Full sun

How to Plant

Starting seeds indoors: Not recommended. Dill has a taproot, which makes it difficult to transplant.

Planting outside: Sow seeds thickly—two seeds per inch—and cover with seed-starting mix. The seeds can take up to two weeks to germinate. Keep the seeds moist while you look for growth.

How to Grow

Water: Keep dill evenly moist.

Fertilizer: No extra fertilizer is needed.

Pest control: Dill is susceptible to the same pests that eat carrots and parsnips. Keep an eye out for aphids and flea beetles. Swallowtail caterpillars, also known as parsleyworms, can be picked off.

When and How to Harvest

The leaves are most fragrant and delicious before the plants start to flower. Use kitchen scissors to snip off leaves. If you want to harvest seeds, allow the plant to flower. As the seeds are forming, keep a close eye on the plants. Pull up the plants and hang them upside down to dry over newspaper to collect the seeds.

GROWING TIPS

Dill gets pretty tall. Stake larger stems to prevent them from falling over in heavy winds or rains.

Don't be alarmed if you see caterpillars on the plants. The native black swallowtail butterfly loves dill and will munch some of your plants to the ground. If you plant dill in several locations in the garden, you might outsmart the caterpillars and have one stand of dill left unscathed.

For a steady supply of fresh, young leaves, sow several successive plantings of dill throughout early spring.

A black swallowtail caterpillar munching on dill.

FENNEL *(Foeniculum vulgare var. azoricum* and *Foeniculum vulgare var. dulce)*

There are two types of fennel: Florence fennel, which is a bulb-forming fennel, and bronze fennel, which is grown primarily for its fragrant and attractive leaves. You can grow bronze fennel all winter long in the warmer regions of the Carolinas. Florence fennel is a spring-planted, summer-growing, fall-harvested plant that does best in container gardens or raised beds in the Carolinas.

■ When and Where to Plant

Temperature: Sow seeds outside when the soil temperature is at least 60 degrees F.

Soil: Fennel needs well-drained, nutrient-rich soil with a pH of 5.5 to 7.0.

Sun: Full sun to partial shade

■ How to Plant

Starting seeds indoors: Not recommended; fennel has a taproot, which makes it difficult to transplant. If you need to start seeds indoors (in the mountains), plant them in a peat or coir pot that can be planted directly outside without disturbing the plant roots.

Planting outside: Sow seeds outdoors, 4 inches apart, and barely cover with seed-starting mix. Thin seeds to 8 inches apart when plants are 4 inches tall. Use the plants you pull up to flavor salads.

■ How to Grow

Water: Fennel is a low-water plant, but you'll need to water it if it doesn't rain at least once a week.

Fertilizer: No extra fertilizer is needed.

Pest control: Fennel is fairly pest-free, but black swallowtail butterfly larvae enjoy munching on it as much as dill and parsley. Let the caterpillars munch. The plant will grow back.

■ When and How to Harvest

Harvest leaves from both types of fennel by using scissors to snip them off at the base of the plant or where they meet the stem. To harvest Florence fennel bulbs, wait until the plant begins to bloom. Then cut off the flowers, wait a few days, and pull up the entire plant.

The leaves of bronze fennel are edible.

GROWING TIPS

For Florence fennel, when the bulb is 2 inches in diameter, start mounding the soil up around it to blanch it (keep it white)

Mulch around fennel plants to keep the roots cool.

PARSLEY *(Petroselinum crispum)*

Parsley is a cool-season, biennial herb grown for its leaves, which are high in vitamins and minerals. In all areas of the Carolinas, parsley will overwinter. It might not produce a lot of new leaves during the very coolest weeks, but it will stay alive, primed and ready to grow. During the second year, parsley produces flowers and seeds, which you can save to plant in fall or the following spring. Parsley is a great addition to cold salads, soups, and sandwiches.

■ *When and Where to Plant*

Temperature: Parsley seeds will germinate with soil temperatures of 50 to 85 degrees F. They grow best in air temperatures of 60 degrees F. You can plant parsley transplants outside from September to October and again from February to April in coastal and Piedmont regions. Plant transplants outside in early April in the mountains.

Soil: Parsley grows best in soils with a pH of 6.5 to 7.5 that are well draining and high in organic matter.

Sun: Full sun to partial shade

Curly parsley

Flat-leaf Italian parsley

How to Plant

Starting seeds indoors: Start seeds indoors at least one month before you want to transplant outside. Parsley can be tricky to start from seed. Soak seeds for twelve hours before planting to hasten germination.

Planting outside: Sow seeds outside at one seed per inch. It can take two to three weeks for parsley seeds to germinate. You can thin the plants to 4-inch spacing once they've germinated.

How to Grow

Water: Parsley won't need extra water unless the area is experiencing a drought.

Fertilizer: Parsley is a heavy feeder. Sidedress with a balanced fertilizer every four weeks during the growing season.

Pest control: The main pest that you'll see on parsley is actually the larval form of the black swallowtail butterfly. These caterpillars will feed on parsley for a few weeks and then disappear.

When and How to Harvest

Keep parsley growing by frequently cutting the outermost leaves for use in the kitchen. You can prolong leaf growth by cutting back the flower stalks for a time. Eventually, though, parsley will flower, set seed, and die.

GROWING TIPS

If you have trouble growing parsley from seed, you can buy transplants locally. Unlike some vegetables and herbs, you don't get much difference in varieties if you don't grow your own from seed.

185

Marigolds serve to ward off pests and attract pollinators in this raised bed garden filled with peppers. Mulch keeps moisture in the soil and helps with weed control.

Warm-Season Gardening Tips & Tricks

The warm-season vegetable gardening season runs from April through September or October in the coastal and Piedmont regions and late May or early June through September in the mountains.

Most of the warm-season vegetables and herbs that we grow are originally from tropical regions. They are tender to frost and don't grow well during cool weather. While they might hang on if temperatures are above freezing but not yet warm, they won't grow.

The biggest mistake that new gardeners make is to plant warm-season vegetables too early. Just because the stores have tomatoes in March doesn't mean you should plant them in March. There is often a late frost right around the time of the vernal equinox and another cold snap right around Easter. Unless you're planning to use a cold frame to elevate the temperature, wait to plant the warm-season vegetables until it is actually warm. That means soil temperatures of 60 to 65 degrees F and nighttime air temperatures of 65 to 70 degrees F.

How Hot Is Too Hot?

There's a flip side to the heat though, and that's extreme heat. Temperatures over 85 degrees F cause plants to slow down and conserve energy. Some plants, such as tomatoes and peppers, can drop their flowers when temperatures are too high. You can mist plants with a fine spray when temperatures soar above 85 or 90 degrees F, but it's better to try to time planting so that the plants grow and set fruit before the dead of summer.

A Second Harvest

Timing is more critical for warm-season gardening than cool-season gardening. Warm-season plants have to be planted after the soil has warmed up so that they can grow to maturity and set fruit before the air becomes stifling. Plants such as tomatoes, summer squash, and cucumbers will grow and fruit and grow and fruit until they, literally, exhaust themselves.

In all but the coldest areas of the Carolinas, you can rip out the first round of these plants and plant another round in early August to mature before the first frost hits. This is noted in individual plant profiles of vegetables and herbs that are good candidates for a second planting.

Watering During the Warm Season

Everything during the warm season is expedited, including water use. You might be able to get away with watering your cool-season vegetable garden once or twice a week, but you will frequently need to water every day or every other day in sandy soils during the warm season. The heat causes plants to use more water and causes water to evaporate from the soil more quickly.

Mulching plants can help cut down on watering frequency. When I started mulching my vegetables, I noticed a huge difference. Soaker hoses and drip irrigation are also helpful tools for the warm-season vegetable garden. Many summer vegetables are sensitive to wet/dry/wet/dry conditions. Mulch and a soaker hose can help keep moisture levels steady to avoid problems such as blossom-end rot and will keep leaves dry, which reduces leaf diseases.

A row of pollinator plants near the vegetable gardens.

Planting for Pollination

Cool-season vegetables are primarily just that—the vegetative parts of the plants. Warm-season vegetables are mostly the fruits of the plants, and to get fruits, flowers have to be pollinated. Corn is a wind-pollinated plant, but it is the exception in the vegetable garden, rather than the rule. Most summer vegetables (fruits) are a result of flowers pollinated by insects. You can increase pollination rates by planting flowers in the vegetable garden.

All plants in the aster family attract pollinators. Lettuce and radishes, if you let a few stay in the garden and flower, will attract hundreds of pollinators. Many herbs, if allowed to flower, will also attract beneficial insects and pollinators. Leave space in the warm-season vegetable garden for flowers. It won't be wasted.

Plants for Pollinators

Basil	Mountain mint	Tarragon
Bee balm	Purple coneflower	Thyme
Lettuce	Radish	Zinnia
Marigold	Sunflower	

Dealing with Pests

Many more pests are active during the warm season than during the cool season. It can be frustrating at times, dealing with so many pests. You'll see that many of the vegetable profiles have recommendations to "use floating row covers" to protect the plants. There are different remedies for pests, listed in the chart on page 65. However, the best two ways to deal with pests have nothing to do with what to spray on the plants.

1. Plant a variety of plants in the garden, and plant groups of the same plant in many places throughout the garden. Diversity is the key to luring beneficial insects and confusing harmful ones. A vegetable garden is not a natural ecosystem, but it can more closely mimic a natural ecosystem if you introduce variety.

2. Raise your threshold for pest damage. I really don't care if most pests eat my plants. The one pest that confounds me and drives me to synthetic chemicals is the squash vine borer. I can't grow squash without chemicals because of this pest. Otherwise, I let the pests munch. I find though, that I have few pest problems, because my gardens are so haphazardly planted and include a mishmash of plants all over the place.

Planting a wide variety of vegetables, herbs, and flowers together confuses pests.

BEAN Bush and Pole Bean *(Phaseolus vulgaris)*
Lima Bean *(Phaseolus lunatus)*

Beans are absolutely warm-weather vegetables. Don't plant these vegetables too early, or you won't get a harvest. There are tons of different varieties of beans, including snap beans—both pole and bush varieties, lima beans, and shelling beans. You can also grow edamame, which are soybeans that you eat while they're still green and tender (but not the pods). Fava beans are the only exception to the warm-weather rule. You plant favas in fall to harvest in spring (in warmer areas) or in early spring for harvest in early summer (in cooler areas). To get a bigger and better bean harvest, inoculate the seeds with nitrogen-fixing bacteria, available online and at natural garden stores.

■ Recommended Varieties

What's your goal when growing beans? Ask yourself that before making selections. Do you have space to grow pole beans on trellises, or do you need to grow more compact bush forms? Do you want to harvest lima beans for making your own succotash or fresh snap beans to lightly steam and eat with summer grilled dinners? Here are some good bean selections for the Carolinas.

Pole beans growing up a trellis.

GROWING TIPS

Don't jump the gun with beans. Wait until the soil has warmed to at least 60 degrees F before planting.

Mulch beans with wheat straw, shredded newspaper, and other low-nitrogen materials to keep moisture levels even and weeds to a minimum.

Place supports before the plants grow taller than 3 or 4 inches. It's difficult to get a trellis in place once the beans start growing. You also want pole beans to have something to "grab" as soon as they're tall enough to start climbing.

Grow beans over your sunny trellis in your flower garden. Scarlet runner beans are beautiful ornamentals as well as edibles, and they can be eaten in the pod when young, eaten fresh and shelled when still tender, or eaten as dried beans.

Baby your beans when the weather is hot. Beans will drop flowers and stop producing when temperatures climb above 85 degrees F. If they can hang on until cooler weather arrives, they'll start producing again. If not, you can pull them up and replant for plants that are maturing when the days have cooled off.

Snap beans, bush type: 'Bush Blue Lake 274', 'Derby', 'Provider', 'Resistant Cherokee Wax' (yellow-podded), 'Roma II' (flat-podded), 'Venture'

Snap beans, pole-type: 'Blue Lake', 'Kentucky Blue', 'Kentucky Wonder', 'Kentucky Wonder 191', 'Kwintus' ('Early Riser'). For something different, plant 'Red Noodle' asparagus beans. They have skinny, 12-inch-long bean pods and are as ornamental as they are edible. Only purchase snap beans that are resistant to bean common mosaic virus and anthracnose—this will be indicated on the seed packet.

Lima beans: You can purchase determinate (bush) and indeterminate (pole) lima beans. The indeterminates are usually thought to have better flavor, but they can grow to be huge and need strong trellises for support. Determinate limas grow to heights of 2 feet and can still use some staking—but this is not as crucial as it is for indeterminate types.

■ *When and Where to Plant*

Temperature: The soil temperature at a 4-inch depth needs to be at least 60 degrees F before planting beans. In coastal and Piedmont regions, you can actually get two bean crops during the year. Plant in mid-March in coastal regions and in mid-April in interior regions for a spring harvest. Plant seeds outside again in July and August for a second harvest. Plant seeds outside in late May for a summer harvest in the mountains.

Soil: Add compost to the soil before planting. Beans will not grow well in soil with a pH below 5.5. Test the pH to ensure it is 6.0 to 7.2.

Sun: Full sun; select an area for your pole beans on the north side of the garden so that they won't shade out the rest of your plants as they grow.

How to Plant

Starting seeds indoors: Not recommended or needed

Planting outside: Plant beans outside when the soil temperature is at least 60 degrees F at a depth of 1 to 1½ inches. You might have heard the term "a hill of beans." That's because beans are usually seeded in mounds or hills, four or five seeds to a hill. If you plant beans this way, you can then put up a trellis or tomato cage over each hill. If you plant beans
to grow along a teepee, plant beans 1 inch deep and 1 inch apart along the bottom of the trellis.

How to Grow

Water: Beans are somewhat drought tolerant until they start flowering. Once plants start flowering, the beans should be kept evenly moist.

Fertilizer: Beans fix nitrogen, so they do not need high-nitrogen fertilizers. Once beans start flowering, fertilize once with a 0-5-5 or 0-10-10 fertilizer if phosphorous and potassium are needed.

Pest control: Pest problems include Mexican bean beetles, thrips, aphids, corn earworms, and stinkbugs. Use insecticidal soap to control thrips and aphids.

When and How to Harvest

Once pods start forming, it's time to harvest. Snap beans should be picked when the pods are still tender. Beans should be easy to snap off the plant and should easily break in half when they're ready to harvest. Pick in the morning when the plants have the highest moisture content. Beans taste best if eaten soon after harvesting, so pick them on the day you plan to eat them. To keep plants producing, pick regularly. If beans are allowed to yellow on the plant, the plant will stop producing more pods.

To harvest limas and pole types for shelling beans, wait until the pods start bulging and then pick. Pick before the pods turn yellow!

To harvest dry beans, allow the pods to dry on the plant. Then pull up the entire plant and hang it upside down to dry over a clean sheet. You will have to thresh the pods in order to remove the dried bits of pod from the dried beans. (Quite frankly, it's easier and cheaper to just buy dried beans.)

Have fun in the garden by making artistic trellises for beans and other climbing vegetables.

CORN *(Zea mays)*

Different types of corn include popcorn, sweet corn, and field corn. This section is about sweet corn. Corn is native to North America. It is in the grass family and started out as a wild grass that was gradually domesticated and bred for larger "ears," which hold the seeds. Even if you have a small garden, you can grow corn. You don't have to have an acre. Corn is wind pollinated, soit is more important to plant corn close together in 4×4- or 6×6-foot blocks than it is to plant a big field.

Corn is a wind-pollinated plant, so you have to grow the plants close together in blocks so the wind can blow the pollen from plant to plant.

■ *Recommended Varieties*

If you have limited space, plant fast-maturing dwarf varieties. 'Earlivee' is one example. Other varieties suited to the Carolinas include:

Yellow: 'Bodacious', 'Merit', 'Kandy Corn'
White: 'Silver Princess', 'Silver Queen'
Bicolor (sometimes called milk and sugar): 'Honey and Pearl', 'Legion'

■ *When and Where to Plant*

Temperature: Sweet corn does not grow well in cold soil. Wait to plant until soil temperatures are 60 to 80 degrees F. This period ranges from mid-March in the southernmost coastal regions to May in the mountains. The best way to check if it's time to plant corn is to use a soil thermometer.

Soil: If you grew peas in spring, plant corn for an early-summer crop. The corn will benefit from the nitrogen fixed by the peas. Add compost to the soil before planting, and work in a balanced, slow-release, organic fertilizer.

Sun: Full sun

■ *How to Plant*

Starting seeds indoors: Not recommended or needed

Planting outside: Wait until the soil has warmed up to plant corn. You can soak seeds in warm water overnight to pre-sprout and ensure germination, but this is not necessary. If you have space to plant 10×10-foot blocks, sow seeds 6 inches apart in rows that are 18 inches apart. If you are growing corn in raised beds, make sure to plant at least twelve stalks spaced 8 inches apart, in a square. You will have to assist with pollination of these plants by shaking them once you see pollen forming.

■ *How to Grow*

Water: Corn requires consistent moisture through the growing season, more moisture when it starts to flower. If it is drought stressed, the leaves will roll up and the plants will take on a "pointy" appearance.

Fertilizer: Corn is a hungry plant. Prepare the planting area by adding compost and a slow-release fertilizer before planting. Fertilize with alfalfa meal and soybean meal when the corn is 2 feet tall and again when it starts to flower.

Pest control: There are many pests that attack corn plants. If you've had problems with earworm, treat ears with a combination of *Bt* and mineral oil (1:20) five days after silks emerge. Place five drops on the silks at the end of each ear.

■ *When and How to Harvest*

Sweet corn is ready to harvest when the juice that squirts from a kernel pierced by a fingernail is milky white. Start checking the corn after the silks (hairs on the ears) start to turn brown. You can hold the corn stalk in one hand and grasp the ear in the other hand, jerking the ear off the plant without yanking the plant out of the ground.

GROWING TIPS

Corn plants are tall and heavy. To prevent plants from falling over in sandy soils, start to hill up the soil around the base of the plant (to a height of 6 inches) when the plants are 1 foot tall.

Cut off any small plants that sprout from the base of the main stalk. These suckers will sap nutrients from the main plant.

Harvest promptly when the corn is ripe. Overripe corn is tough and chewy.

CUCUMBER *(Cucumis sativus)*

Cucumbers are warm-season vining vegetables that are part of the squash family. Sometimes you'll hear members of the squash family called "cucurbits." Just as the cabbage family includes many different but related varieties of cold-season vegetables, the cucurbit family includes many related vegetables. This presents somewhat of a challenge, as some of the pests that attack one type of plant in the family can also attack the other types as well. Cucumber beetles, squash vine borers (which do not attack cucumbers), and several other pests find the summer garden to be an all-you-can-eat buffet of squash-family plants. That's no reason not to grow them, though. Here's how to have success with cucumbers.

■ *Recommended Varieties*

There are several types of cucumbers. Slicers are larger and longer cucumbers made for, you guessed it, slicing. Picklers are shorter, smaller cucumbers. Burpless varieties have thinner skin. There are also vining types and bush types. Bush types are better if you have limited space. Vining types need heavy-duty support in order to produce well. Here are some recommended varieties for the Carolinas:

Slicers: 'Salad Bush' (hybrid), 'Straight Eight', 'Sweet Slice', 'Sweet Success' (hybrid), 'Burpless' (hybrid), 'Poinsett 76'

Picklers: 'Fancipak' (hybrid), 'Calypso', 'Carolina', 'County Fair', 'Homemade Pickles', 'Excusion'

In addition to considering the growth habit of the plant, when shopping for cucumber seeds also look at disease resistance and try to find varieties

with resistance to all or some of the most prevalent cucumber diseases. If you see CMV on the seed packet, it means the cucumber variety is resistant to cucumber mosaic virus. DM means resistance to downy mildew, and PM means resistant to powdery mildew. DM resistance is very important for season-long cucumber production.

Growing cucumbers up a trellis saves space in the garden.

When and Where to Plant

Temperature: You can't plant cucumbers until all danger of frost has passed and the soil is at least 60 degrees F. A soil temperature of 70 degrees F is even better. In the Carolinas, you can plant twice—in spring and early in fall to get two harvests. You can plant outside around March 15 in the southernmost regions, in mid to late April in northern coastal regions, in mid-April to mid-May in the Piedmont, and in late May to early June in the mountains. Plant a second round of seeds outside for a fall harvest in all areas around August 1.

Soil: Cucumbers grow best in soils with a pH of 5.5 to 6.8. They are also heavy feeders and use a lot of water, but they need well-draining soil. Add compost to the soil before planting in order to grow healthy plants.

Sun: Full sun

How to Plant

Starting seeds indoors: Not recommended; cucumbers and other squash plants do not transplant well. They also sprout so quickly that you don't gain anything by starting seeds inside.

Planting outside: Sow seeds outside when the soil temperature is at least 60 degrees F and all danger of frost has passed. Allow 8 to 12 inches between plants.

How to Grow

Water: Cucumbers are heavy drinkers. Keep the soil evenly moist during the entire growing period. These plants need more water than almost any other vegetable. Place soaker hoses along the base of cucumber plants to help deliver a steady stream of water at the roots.

Fertilizer: Cucumbers need a lot! Sidedress cucumbers every two weeks during the growing season with a balanced fertilizer.

Pest control: Cucumber beetles and pickleworms are the peskiest pests to attack cucumbers. Neem oil and pyrethrum are two organic pesticides that can help control the pests.

When and How to Harvest

Once cucumber fruits start forming, keep your eyes peeled for fruits that needs to be picked. Fruits left unpicked will also stimulate the plant to stop fruiting. Harvest pickling cucumbers when they are 2 to 6 inches long. Harvest slicing cucumbers when they are 6 to 12 inches long. To see how long the cucumbers you're growing should be at maturity, consult the seed packet. Use pruners to cut the fruits off, leaving 1 inch of stem on the end of the fruit. (This minimizes water loss after harvest.)

EGGPLANT *(Solanum melongena)*

Eggplants, along with tomatoes, peppers, and tomatillos, are members of the nightshade family. If the cucurbit family claims half of the warm-season vegetables we grow, the nightshade family owns the rest. All nightshade vegetables are originally from warmer, tropical locations, and, as such, do not grow well in the cold. They do, however, grow well in summer in the Carolinas. I've even successfully grown eggplants in a bit of afternoon shade. At a loss for what to do with eggplants? Two easy ways to eat them are rubbed with olive oil and grilled, and in a ratatouille (summer sauté of tomatoes, squash, and eggplant). If you have the time and patience, make your own eggplant parmesan or lasagna. It will taste so much better than what you can get at a store or restaurant.

Recommended Varieties

Explore varieties beyond just the "regular" bowling-pin-sized eggplants. There are eggplants that produce long, skinny, dark fruits; light purple fruits; lavender fruits; and white fruits. Here are some varieties to try:

Regular, large, dark purple type: 'Black Beauty', 'Purple Rain', 'Classic'

Oriental type (long, skinny fruits): 'Little Fingers', 'Orient Express', 'Ichiban', 'Pingtung Long'

Other types: 'Casper' (white fruits), Italian Pink Bicolor, also known as 'Rosa Bianca' (large bell-shaped fruit with pink stripes)

When and Where to Plant

Temperature: Eggplants need soil and nighttime air temperatures to be at least 65 degrees F to grow. You can plant outside in early April in southern coastal regions, mid to late April in northern coastal regions, early to mid May in the Piedmont, and late May to early June in the mountains.

GROWING TIPS

Eggplants (the fruits) are heavy. Stake plants individually (one stake per plant, placed next to the stem and tied) to support the plants and keep the fruits from sitting on the soil.

If you've had problems with insects, one way to thwart them is to plant groups of the same vegetable in different areas of the vegetable garden. The pests might find one group, but will leave the others alone.

Soil: The pH should be 5.0 to 6.5 for best growth. Add compost to the soil before planting, and plant in well-drained locations.

Sun: Full sun to partial shade (needs full sun in mountains)

How to Plant

Starting seeds indoors: If you want to grow your own transplants, start seeds indoors six to eight weeks before you want to plant outside. Use a heat mat and grow lights for strong plants.

Planting outside: Plant transplants outside after all threat of frost has passed, with 12 to 18 inches between plants. (You can get away with closer spacing in raised beds—12-inch centers work in these.)

Eggplants come in different shapes and sizes. Look at your seed packet to tell what they're supposed to look like when ripe so you know when to harvest.

How to Grow

Water: Eggplants need a large, consistent supply of water. Use soaker hoses around plants for the best result.

Fertilizer: Feed eggplants every two weeks with a balanced fertilizer. They like to eat.

Pest control: Many pests and diseases call eggplants home. Colorado potato beetles, tomato hornworms, and flea beetles are the worst insect offenders. Keep an eye out for tomato hornworms and handpick them off. (Gross, but effective.) Look for the orange eggs of the Colorado potato beetle on the undersides of the leaves, and wash the leaves with water or crush the eggs to control them. Specific strains of *Bacillus thuringiensis* (*B.t.*) are effective on adult beetles.

When and How to Harvest

The biggest question with eggplants is, when are they ready to harvest? To know the right answer, you need to pay attention to the seed packet or plant label. Some eggplants grow to be 6 or 8 inches long, while others max out at 4 inches. If you leave the fruits on the plant too long, they'll grow big seeds and become bitter tasting. Ripe eggplants have shiny skin. If the skin is dull, they're past their prime, and you should pick them and throw them in the worm bin or compost heap. Use a knife or pruners to pick the fruits, leaving at least ½ inch of stem.

OKRA *(Abelmoschus esculentus)*

Okra is a hot-weather vegetable known for its pointy pods that have a somewhat slimy texture when cooked. Although that may not sound very appetizing, okra is actually really delicious if prepared correctly. It's a key ingredient in gumbo but also tastes wonderful when brushed with olive oil and roasted or dipped in cornmeal and fried. It's easy to find frozen okra, which is fine for soups, but you'll want fresh okra for roasting or frying. Do you like to make your own pickles? Pickled okra is always the first thing to disappear off of any crudité plate I put out at an event. In addition to its versatility, the flowers of this vegetable in the mallow (hibiscus) family are pretty, so make room in the garden for it.

■ *Recommended Varieties*

'Clemson Spineless', 'Lee', and 'Annie Oakley II' are all spineless varieties that grow well in the Carolinas. 'Prelude' and 'Emerald' are open-pollinated varieties (you can save the seeds). 'Prelude' has longer pods than most hybrids. 'Red Burgundy' has red, not green, pods.

■ *When and Where to Plant*

Temperature: Okra likes it hot. Don't bother planting until the soil temperature is at least 65 to 70 degrees F at a depth of 4 inches. That ranges from early April in the southernmost areas of the Carolinas to early June in the mountains. Keep your soil thermometer handy!

GROWING TIPS

While few pests attack okra, there are some diseases that will take hold if the leaves of the plants stay wet and there's low air movement. To get a good harvest, allow enough space between plants for airflow. Even in raised beds, you can't cram these plants together.

Okra needs well-drained, loose soil to thrive. If you have a problem with clay, consider growing okra in a container or in a raised bed that drains well.

Because okra grows so tall, you might need to stake the plants. They'll keep growing and producing best if they stay upright, so plant them along a trellis or stake each plant individually.

In the warmest areas, you can cut okra plants back by half in mid- to late summer. They'll grow out and produce a second fall crop.

Soil: Rotate the planting location of okra from year to year to keep plants from suffering from soilborne diseases.

Sun: Full sun; plant on the north side of the garden so that the tall plants don't shade other plants reaching for the sun.

How to Plant

Starting seeds indoors: Not necessary; but it is helpful to soak the seeds for twelve hours before planting outside to ensure good germination rates.

Planting outside: If planting transplants, try not to disturb the roots—okra is finicky about transplanting. Use scissors to cut the pot away from the plant, rather than yanking the plant out of the pot. Space transplants 9 to 12 inches apart. Sow seeds at a rate of one every 4 inches. Thin to 9- to 12-inch spacing after plants have germinated and have three sets of leaves.

How to Grow

Water: Allow the soil to dry out between plantings. Water deeply (so that the soil is wet to a depth of 6 to 8 inches) but infrequently.

Fertilizer: Fertilize once during the growing season with a balanced fertilizer.

Pest control: One reason to grow okra: no bad pest problems.

When and How to Harvest

Harvest okra when the pods are 2 to 3 inches long. The larger the pod, the tougher it will be, and the less tasty. Once an okra plant starts producing pods big enough to harvest, it will keep going until frost occurs or it wears itself out. Plan to pick new pods every single day. If you miss pods and they get too large, pick them off and compost them—leaving them on will signal the plant to stop producing.

Okra fresh from the garden is best for roasting or frying.

The Carolinas are a great place to grow peppers.

PEPPER *(Capsicum annuum)*

Like eggplants and tomatoes, peppers are members of the nightshade family. The only part of the plant that is edible is the fruit. Never eat the flowers or the leaves—they're poisonous. (Not that you probably would, but if there is one plant that you don't want to experiment with, it's peppers.) There are hundreds of varieties to grow—from bell peppers to banana peppers, from hot to sweet. As with all of the warm-season vegetables, almost every variety of pepper will grow equally well in the Carolinas. Regions that don't have enough heat hours can struggle getting peppers to fruit, but that is definitely not something we have to worry about here! So pick a potted pepper and plant it. You're almost guaranteed success. It's expensive to buy fresh peppers at the grocery, so learning how to grow these vegetables can have a significant impact on your shopping bill.

■ *Recommended Varieties*

Any variety you select will grow, as long as you give it the right conditions. Always look for disease resistance when selecting plant types. This will be indicated on the plant tag or seed packet. Here are some favorites to try:

Bell pepper: 'Keystone Resistant Giant Strain 3', 'Yolo Wonder I', 'King Arthur'

Banana pepper: 'Banana Supreme', 'Hy-Fry', 'Biscayne', 'Key Largo', 'Cubanelle', 'Gypsy', 'Hungarian Sweet Wax'

Hot pepper: 'Hot Cherry', 'Anaheim Chili TMR 23', 'AnchoVilla', 'Early Jalapeno', 'Mitla', 'Hungarian Yellow Wax', 'Habanero'

Pimento: Pimento Select, True Heart Perfection

■ *When and Where to Plant*

Temperature: Peppers need warm soil to grow—at least 65 to 75 degrees F at a depth of 4 inches. They will experience blossom drop (and lowered fruit production) if daytime temperatures are consistently over 95 degrees F. Plant transplant peppers outside in early April in southern coastal locations, late April in northern coastal locations, May in the Piedmont, and June in the mountains. Do not plant outside until all danger of frost has passed.

Soil: Peppers grow best with a soil pH of at least 6.0. Bell and sweet peppers are heavy feeders, so add compost to the soil before planting.

Sun: Full sun

■ *When and How to Plant*

Starting seeds indoors: It's advantageous to start seeds indoors yourself or purchase transplants so that you end up with productive plants before the real heat of summer bears down on the area. Start seeds indoors six weeks before you plan to transplant outside. You want the transplants to be at least 6 inches tall before planting outside. For strong transplants, use a heat mat placed under the seedling flat and grow lights 2 to 4 inches above the flat. (It is helpful to put grow lights on chains so that you can easily raise them as the plants grow.

Leave bell peppers on the plant longer for a red or orange color.

To get strong seedlings, follow this protocol: Sow seeds in a seedling flat with a heat mat under it. When plants sprout and have their first leaves, leave the grow lights on the plants for at least sixteen hours per day. When the first true leaves (second set of leaves) appear, transplant seedlings into 4-inch pots. Let the plants grow. Harden off the plants when all danger of frost has passed, and then plant in the garden.

Planting outside: Plant peppers outside when soil temperatures are at least 65 degrees F. Plant transplants 12 inches apart.

▇ *How to Grow*

Water: Bell and sweet peppers thrive with regular water. If they go through wet/dry/wet/dry cycles and/or the pH of the soil is too low, they develop blossom-end rot, a condition in which the end of the fruit opposite the stem rots. Hot peppers are one of the most drought-tolerant vegetables but benefit from watering when first planted.

Fertilizer: Feed bell and sweet peppers with a balanced fertilizer when they start blooming and again four weeks later.

Pest control: Peppers are relatively pest-free, specially hot peppers.

Banana peppers are easy to grow!

GROWING TIPS

Use soaker hoses around pepper plants so you can easily apply consistent moisture.

Bell and sweet peppers thrive when mulched with shredded newspaper, shredded hardwood, and other organic mulches.

Stake plants individually when they grow to heights of 18 inches or taller. Wear gloves when harvesting hot peppers.

◼ *When and How to Harvest*

Read the seed packet or plant tag to see when your peppers will be ready to harvest. There will be an indication of days to maturity (after transplanting outside), as well as pictures that show you what the "finished product" looks like. Some red bell peppers are green when they're immature and turn red eventually. Some bell peppers never turn red. Hungarian wax peppers can be picked when green or red. It gets confusing. The longer peppers grow, the sweeter or hotter they get.

SUMMER SQUASH *(Cucurbita pepo)*

If you grow more than two or three summer squash plants, you will have to engage in what I like to call the "zucchini drop and dash." This is when you leave bags of summer squash on porches, in unlocked cars, and anywhere anyone isn't looking, really. These plants are prolific if you can keep the squash vine borers away from them. They also grow really fast. Forget to pick a zucchini one day, and you're playing squash bowling the next day with the giant bowling-pin-sized fruits. Luckily there are as many ways to eat summer squash as there are fruits produced on each plant.

Recommended Varieties

Summer squashes include all of the thin-skinned squashes that are suitable for eating cooked or raw. Yellow crookneck varieties are popular, as are typical dark green zucchini plants. You can grow pattypan squash, which look like little yellow and white space ships, or round zucchini that look like pool balls. Try these varieties:

Yellow straightneck: 'Saffron', 'Seneca', 'Butterbar', 'Multipik'
Yellow crookneck: 'Dixie', 'Yellow Crookneck'
Zucchini: 'Embassy', 'Spineless Beauty', 'Senator'
Pattypan: 'Sunburst', 'Peter Pan'

When possible, read the seed packets to purchase varieties that are resistant to cucumber mosaic virus, powdery mildew, and downy mildew.

Harvest summer squash right after the flower on the end of the fruit wilts.

When and Where to Plant

Temperature: Squashes need warm soil to grow. Plant outside when soil temperatures are at least 65 degrees F. In southern coastal areas, plant seeds outside from March 20 to April 15 and again August 15 to 30. Plant in northern coastal areas from April 15 to May 1 and again August 15 to 30. In the Piedmont, plant outside from May 1 to June 1, and in the mountains, plant outside June 1 to 30.

Soil: Squashes grow best in a pH of 6.0 to 6.5. Add compost to the soil before planting. If the soil test shows low boron levels, scratch 1 teaspoon of borax into the soil per plant.

Sun: Full sun

How to Plant

Starting seeds indoors: Squash plants are difficult to transplant. If you want to get a jump on the season, start seeds indoors two weeks before you want to plant outdoors. Plant the seeds in peat pots or CowPots that you can plant directly in the ground without removing the plants.

Planting outside: Sow seeds outside when soil temperatures are at least 65 degrees F. Plant two to three seeds per hole, 1 inch deep. When seedlings are 3 inches tall, use scissors to thin to two plants per 12 inches. As plants grow, you can further reduce to one plant per 18 or 24 inches. (Overplanting slightly gives you more plants to work with if you have problems with pests.)

How to Grow

Water: Keep squashes evenly moist. It helps if you mulch the garden after the plants germinate.

Fertilizer: Sidedress with a balanced fertilizer every three weeks during the growing season.

Pest control: Squash bugs, squash vine borers, cucumber beetles, and squash beetles plague squash plants. Use row covers when plants are young to protect plants from the flying pests. You can raise the covers for two hours in the early morning a few days a week to allow insects to reach the flowers for pollination.

When and How to Harvest

Once summer squash plants start producing, you need to check the plants daily for fruits to harvest. Summer squash tastes better when harvested young. You can even harvest with the flowers still hanging on to the ends of the fruits.

Sweet potatoes grow best in the long growing season and warm weather of the Carolinas.

SWEET POTATO *(Ipomoea batatas)*

Not everyone can grow sweet potatoes. They are warm-season plants that are highly sensitive to cold weather and require a long frost-free growing season. Drive through the Carolinas in fall and you'll see fields of sweet potatoes that have just been dug and are drying before being taken to storage. Sweet potatoes are easy to grow in well-drained, sandy soil, once established. Dedicate one of your raised beds or a corner of your garden to growing your own sweet potatoes. You can basically plant them and forget about them, and dig up a bountiful harvest to last through the winter.

Recommended Varieties

Look for these varieties of sweet potatoes that are disease resistant or resistant to nematodes that can be problematic with sweet potatoes.

Beauregard: Resistant to diseases, but not nematodes
Centennial: Resistant to root knot nematode and wireworm
Excel: Resistant to southern root knot, stem rot or wilt disease, and internal cork
Jewel: Resistant to wilt and root knot nematode
Regal: A great variety for home gardeners, because it is resistant to almost every sweet potato problem and is a true "plant and forget" variety

When and Where to Plant

Temperature: Plant slips outside when the soil temperature is at least 65 degrees F. Planting times range from early May in southern coastal regions to early June in the Piedmont. Mountainous regions might not have enough frost-free days to grow sweet potatoes. Consult the frost tables for your area before planting sweet potatoes.

Soil: Sweet potatoes need soil that is slightly acidic (pH of 5.5 to 6.0), sandy, and fast draining. If you have heavier clay soil, plant slips in raised beds.

Sun: Full sun

How to Plant

Starting indoors: You don't grow sweet potatoes from seed—you grow them from slips, or sprouted roots from other sweet potatoes. You can also buy transplants.

Planting outside: Plant slips or transplants outside when the soil temperature is at least 65 degrees F. Sweet potatoes really need hot weather, or they'll just sit in the ground. Plant slips 4 inches deep and 12 inches apart in rows 36 inches apart.

How to Grow

Water: Water while plants are establishing themselves (the first couple of weeks). After that, the plants will not need much extra water unless there is no rain for more than a week at time.

Fertilizer: Sweet potatoes are not heavy feeders. You can sidedress with a low-nitrogen, high-phosphorus fertilizer four weeks after planting.

Pest control: Sweet potatoes are largely pest-free, especially if you plant resistant varieties. If you grow sweet potatoes each year, you can keep problems at bay by rotating where you plant them each year. Deer love sweet potato and can be a major pest. In areas with high deer pressure, fencing may be necessary.

When and How to Harvest

Sweet potatoes are ready to harvest about 120 days after planting or when the roots are at least 3 inches in diameter. Cut and remove the vines to compost, and then gently dig around the plants and pull up the roots. Allow the roots to dry on top of the soil for a day. After curing for a day outside, pick up the roots and bring them inside. Cure the potatoes inside for a week before putting them in storage. Set the potatoes out in a warm, dark room (85 to 90 degrees F) with high humidity. This allows any cuts or bruises to heal. Then move them to a cool (55 to 65 degrees F), dark area and leave them alone until you're ready to use them.

GROWING TIPS

Sweet potatoes will rot if given too much extra water. Once the plants have become established, don't give them extra water unless you go a week without rain.

It can be hard to find sweet potato slips. Check farm stores in your area or order slips online.

If you can, grow sweet potatoes in raised beds. They warm up earlier in spring, stay warm longer, and drain quickly.

TOMATILLO *(Physalis ixocarpa)*

Sometimes known as the "husk tomato," tomatillos are green fruits that grow inside a papery shell. You can use them fresh, roasted, or sautéed. Tomatillos have a slightly tart flavor, but cooking them brings out some sweetness. A few tomatillo seeds go a long way because tomatillos are ridiculously easy to grow from seed. In fact, they're as easy to grow from seed as from transplants. They are also prolific producers that reseed all over the place.

Recommended Varieties

'Toma Verde' is a smaller, sweeter variety of tomatillo. 'De Milpa' is a lavender-colored heirloom tomatillo. 'Purple' is a variety that is purple inside and out. Don't get overly concerned with selecting a variety, though. Plant a tomatillo seed, and it will grow.

Wait until the husks turn brown and papery before harvesting tomatillos.

When and Where to Plant

Temperature: Tomatillos like it hot. Plant seeds (or transplants) outside when the soil temperature is at least 60 degrees F. This ranges from mid-March in the warmest areas of our region to late May in the mountains.

Soil: Tomatillos aren't picky plants but will grow best in well-drained soil. If you have clay soil, amend with compost before planting.

Sun: Full sun

How to Plant

Starting seeds indoors: You only need to plant tomatillo seeds indoors if you grow them in the cooler mountainous regions. Start seeds inside six weeks before you plan to plant outside. For strong seedlings, use a heat mat under the seedling flat and grow lights 2 inches above the plants (moved up as they grow).

Planting outside: Plant seeds outside when the soil temperature is at least 60 degrees F. Cover lightly with seed-starting mix, and keep moist. When the plants are 3 inches tall, thin to allow 12 inches between plants.

How to Grow

Water: Keep tomatillos evenly moist.

Fertilizer: Sidedress with a balanced fertilizer four weeks after planting.

Pest control: Keep an eye out for pests that attack tomatoes, such as tomato hornworms, aphids, and cutworms. You can protect plants from cutworms by loosely wrapping the bottom 3 inches of the stems with newspaper.

When and How to Harvest

Pick tomatillos off the plants when the husks start to dry out and turn light brown.

GROWING TIPS

Place tomato cages around individual plants to keep them from flopping over. For a bushier plant, pinch off the top of the plant to encourage side sprouts.

If growing tomatillos from seed, hill the soil up around the bottom 6 inches of the stem as the plant grows. If planting transplants, plant 4 or 5 inches deep to encourage rooting from the stem.

TOMATO *(Lycopersicon esculentum)*

Nothing beats the taste of a fresh-picked tomato from the garden. I grew up not even liking tomatoes until my eighth-grade teacher dropped off a bag at my house for me to try. (We became gardening friends after I "graduated" middle school.) I was hooked. Today's commercial tomatoes are grown for their ability to withstand shipping, not for their flavor. Tomatoes that you purchase in the grocery store have been picked when still green and gassed with the plant hormone ethylene to get them to turn red. No wonder I didn't like them. A farmers' market tomato is closer to the "real thing," but if you want to can, dry, or cook with massive amounts of fresh tomatoes, the only thing to do is grow them yourself.

A prolific tomato plant can produce almost more fruits than you can use. However, there are so many pests and diseases that plague tomatoes that it is always a good idea to overplant. You can always find a home for fresh garden tomatoes if you have too many.

■ *Recommended Varieties*

There are three main types of tomatoes available. Patio tomatoes are small, basically dwarf, tomatoes that grow in raised beds or containers. Most patio tomatoes have small cherry- or grape-sized fruits. Determinate tomatoes are still large vines, but they eventually stop growing taller and just grow more fruits. Indeterminate tomatoes are vining plants that will grow forever if not cut back. Wild tomatoes are vining plants that can grow 10 feet or more!

The key to success with tomatoes is to start with varieties that have a proven track record in the Carolinas. Hot, humid summer weather is challenging for plants. It provides a wonderful environment for pests and diseases to flourish. Most of the tomato varieties that grow well here are resistant to some of the common diseases that plague tomatoes.

Read the seed packets and plant labels to determine disease resistance. Plants resistant to fusarium wilt will have F, FF, or FFF after the cultivar name. VFN on the label indicates that the plants are resistant to verticillium wilt, fusarium wilt, and root knot nematodes; VFNT indicates resistance to these three diseases, plus tobacco mosaic virus. Plants resistant to tomato spotted wilt virus have TSWV on the plant tags. Resistance to root knot nematodes and TSWV is particularly important in the coastal plain.

High-performing tomatoes for the Carolinas include 'Better Boy', 'Whopper', 'Celebrity', and 'Mountain Pride'. I've also had success with 'German Giant', but this plant does live up to its name. It's huge!

'Yellow Pear' tomatoes are indeterminate tomatoes that produce tons of yellow, lightbulb-shaped grape tomatoes. Two of these plants will give you enough tomatoes for every salad you eat all summer long.

Check at your local farmers' markets to see what tomatoes those growers recommend. They'll have experience growing for market in your area.

When and Where to Plant

Temperature: Tomatoes are absolute heat lovers. Don't bother planting until the soil temperature is at least 65 degrees F and nighttime temperatures are regularly at least 65 to 70 degrees F. Late-spring frosts around the area can dash the hopes of many a tomato lover's dream. Plant outside from early April in the southernmost coastal region to early June in the mountains.

Soil: The soil pH should be 6.0 to 7.0. If you have more acidic soil, add lime before planting. Mix in slow-release, organic fertilizer such as Plant-tone before planting, as well.

Sun: Full sun

How to Plant

Starting seeds indoors: Start seeds indoors six weeks before you want to plant tomatoes outdoors. Use a heat mat and grow lights elevated 2 inches above the plants to grow strong transplants. Harden off plants before planting them in the garden.

Planting outside: You can actually direct-sow tomatoes in the Carolinas. That sounds strange, but I've done it and harvested tomatoes at the same time I got fruit from my transplants. If you want to try direct-sowing, sow seeds thickly on top of the soil and cover with seed-starting mix. Keep the

For best results, keep tomatoes consistently and evenly watered.

213

GROWING TIPS

Place tomato cages and supports when you plant the plants. It is easier to do this when the plants are small than after they start growing. You can also stake individual plants with metal or wood stakes placed next to the stem and tied to the plant.

Chipmunks and squirrels will take bites out of tomatoes if they're having a hard time finding water. You can spray tomatoes with repellents to protect fruits.

Pinch suckers (sprouts growing from between the main stem and productive branches) to direct resources to the main fruiting branches.

It's difficult to control many pests, but you can handpick tomato hornworms, the fat green worms that like to munch on the plants.

Yellow pear tomatoes are prolific producers.

Cracking fruit is caused by uneven watering. Use soaker hoses with tomatoes to avoid this problem.

seeds moist while sprouting—you might have to water twice a day. As the plants start growing, thin so that plants are spaced 18 to 24 inches apart. Place cages and supports once the plants are 6 inches tall. Hill up soil around the bottom 6 inches of each plant stem to encourage strong roots. (Patio tomatoes don't have to be hilled or staked.)

Plant transplants outside after hardening off. Space plants 18 inches apart, and plant transplants 5 inches deep to encourage rooting from the stem and a deep root system.

How to Grow

Water: Tomatoes need steady, consistent moisture in order to avoid blossom-end rot (the end of the fruit opposite from the stem rots), a cultural problem that arises when calcium is deficient in the soil or watering is inconsistent. Use soaker hoses to make it easier to water tomatoes deeply.

Fertilizer: Avoid high-nitrogen fertilizer, which can cause more leaf growth than flower development. Fertilizers containing potassium can improve flavor and disease resistance.

Pest control: Almost every pest alive enjoys eating tomatoes, including deer. Tomatoes are also attacked by many diseases. Growing disease-resistant varieties helps avoid many disease problems.

When and How to Harvest

Tomatoes are ripe when you need little effort to pull them off the plant. Leave tomatoes on the plant for as long as possible for the sweetest fruits.

WINTER SQUASH *(Cucurbita* spp.*)*

Winter squash grows in much the same way as summer squash, but the fruits take longer to mature. They also keep longer in storage. Popular winter squash varieties include pumpkins, acorn squash, and butternut squash. The term "winter squash" is a bit of a misnomer here, as winter squashes are ready for harvest between July and September, depending on when they were planted. You do not need to leave them in the garden until the weather cools off.

Recommended Varieties

Winter squash: Varieties that do well in the Carolinas include Early Butternut, Cornell's Bush Delicata, and Table Ace.

Pumpkins: Pumpkins are divided into different categories, including carving (large), pie (small and sweet), and gourds or decorative. Try these varieties:

Carving: 'Autumn Gold', 'Sorceror', 'Magician', 'Howden', 'Magic Lantern', 'Merlin', 'Spirit'

Pie: 'Amish Pie', 'Small Sugar'

Giant: 'Big Max', 'Big Moon', 'Prizewinner'

Novelty: 'Baby Boo', 'Jack Be Little', 'Lumina'

When and Where to Plant

Temperature: Squashes need warm soil to grow. Plant outside when soil temperatures are at least 65 degrees F. In southern coastal areas, plant seeds outside from March 20 to April 15 and again August 15 to 30. Plant in northern coastal areas from April 15 to May 1 and again August 15 to 30. In the Piedmont, plant outside from May 1 to June 1, and in the mountains, plant outside June 1 to 30.

GROWING TIPS

Keep fruits from rotting while they grow by raising them up off the ground on planting flats or crates.

Mulch around the plants to keep the roots cool, keep the soil moist, and minimize weeds.

Winter squash plants are big and need a lot of room. If you have the inclination, you can build a sturdy trellis from pipe and wire and train them to climb. If you do this, you might need to support the fruits in homemade slings.

Soil: Squashes grow best in a pH of 6.0 to 7.5. Add compost to the soil before planting. If the soil test shows low boron levels, scratch 1 teaspoon of borax into the soil per plant.

Sun: Full sun

How to Plant

Starting seeds indoors: Squash plants are difficult to transplant. If you want to get a jump on the season, start seeds indoors two weeks before you want to plant outdoors. Plant the seeds in peat pots or CowPots that you can plant directly in the ground without removing the plants.

Winter squash can be grown as an edible or as a decorative gourd.

Planting outside: Sow seeds outside when soil temperatures are at least 65 degrees F. Plant two to three seeds per hole, 1 inch deep. When seedlings are 3 inches tall, use scissors to thin to two plants per 12 inches. As plants grow, you can further reduce to one plant per 24 or 48 inches. (Overplanting slightly gives you more plants to work with if you have problems with pests.)

How to Grow

Water: Keep squashes evenly moist. It helps if you mulch the garden after the plants germinate.

Fertilizer: Sidedress with a balanced fertilizer every three weeks during the growing season. Apply calcium nitrate at a rate of 2 pounds per 100 feet of row to the soil around the base of each plant when the plants start flowering.

Pest control: Squash bugs, squash vine borers, cucumber beetles, and squash beetles plague squash plants.

When and How to Harvest

Winter squashes are ready to harvest when the skin has thickened and is dull, not shiny. Cut squashes from the vine, leaving 1 inch of stem. Wipe the soil off the fruits, and then use a clean cloth dipped in chlorine or a dilute bleach solution to wipe them down to kill any bacteria on the outside.

Basil is a warm season herb that's easy to grow.

Warm-Season Herbs

Many of the herbs that flourish during warm weather are also cold-hardy perennials, but they don't grow and thrive during the cooler weather. For all of these herbs to produce fresh growth that's ideal for cooking, they need full sun and warm temperatures.

Some warm-season herbs are perennials, while others are annuals. Perennial herbs are ideal for planting in the flower garden or right outside your kitchen door. I find that I use what's near the kitchen more frequently than what's out in the "back 40" where my vegetables are planted. I do find that it's easier to maintain annual herbs if they're growing with the vegetables. The annual warm-season herbs usually need more water than the perennials.

Culantro is a good summer substitute for cilantro.

What about Salsa Gardening?

Cilantro, everyone's favorite herb for salsa, is actually a cool-season herb. So by the time your tomatoes ripen, your cilantro is usually long gone. However, if you plant a late crop of cilantro and baby along an early crop of tomatoes, it is possible for these two crops to overlap in your garden. If you can't grow cilantro in summer, try growing culantro (*Eryngium foetidum*). It has the same flavor but thrives in warm weather.

Cut off basil flowers to keep the plants producing fragrant, tender leaves.

BASIL *(Ocimum basilicum)*

What's a fresh garden tomato without basil? Some people love making their own pesto from bunches of this prolific herb. While pesto isn't my cup of tea, I love basil with tomatoes and eggplants, with squash, in salads, or on toast with fresh mozzarella. Basil is a perfect complement to warm-season vegetables. It's a true warm-season herb that thrives in hot weather. It's easy to grow and comes in a variety of different flavors and varieties.

▇ *Recommended Varieties*

There are so many varieties of basil. All will grow equally well if planted when it is hot outside. Try lemon basil for a different flavor in cocktails and lemonade. 'Genovese' is the classic pesto basil variety. You can also find 'Lime' basil and 'Cinnamon' basil. 'Italian Large Leaf' basil is a good all-around variety. 'African Blue' is a fairly cold-tolerant basil that you can grow on a sunny windowsill in the garage during winter. 'Greek Columnar' is a nonflowering variety of basil.

GROWING TIPS

Cut the flowers off basil to encourage more production of fragrant leaves.

If basil plants get too leggy, just chop the plant back by two-thirds to one half, water, and fertilize well.

When and Where to Plant

Temperature: Plant outside when nighttime temperatures are consistently at least 70 degrees F. You can plant basil outside in the southernmost areas in mid-April to late May or in early June in the mountains. Plant basil at the same time you plant your tomatoes.

Soil: Basil grows best in soil with a pH of 5.5 to 7.0. Add compost to the soil before planting.

Sun: Full sun

How to Plant

Starting seeds indoors: Start seeds indoors two weeks before planting outdoors.

Planting outside: Plant transplants outside, 12 inches apart.

How to Grow

Water: Basil isn't a heavy drinker, but keep it evenly moist.

Fertilizer: As with most herbs, basil doesn't require a lot of extra fertilizer. In fact, extra fertilizer will make basil less fragrant.

Pest control: Japanese beetles and slugs are the most problematic pests for basil. There's not much you can do about Japanese beetles, but you can spread diatomaceous earth around the base of each plant to keep slugs from munching.

When and How to Harvest

Snip leaves from the top of the plant in the morning. Basil is a forgiving and fast-growing plant. You can chop off any piece of the plant and it will regrow.

BAY *(Laurus nobilis)*

Bay laurel is actually a small tree that is hardy outdoors in coastal areas (zone 8). If you grow only one herb, grow this. Plant it in full sun where it has room to grow to a height of at least 5 feet, and leave it alone. If you can plant it near the kitchen door, you'll be more likely to run out and pick a few leaves for soups and roasts. Fresh bay leaves are so much better to cook with than dried leaves. This herb could just as easily be categorized as a cool-season herb. It's evergreen, so you'll have fresh leaves available for use all year long.

Recommended Varieties

It can be hard to find this plant. Plant whatever you can locate!

When and Where to Plant

Temperature: Plant outside at any time of year in coastal areas.
Soil: Bay laurels prefer well-drained soil.
Sun: Full sun

How to Plant

Starting seeds indoors: Not recommended
Planting outside: Plant one bay laurel plant outside in full sun where the plant can branch out and grow to a height of at least 5 feet.

How to Grow

Water: Bay laurels have low water needs once established.
Fertilizer: No fertilizer is needed.
Pest control: Bay is not susceptible to pests.

When and How to Harvest

Pick off individual leaves for use in cooking.

GROWING TIPS

Bay laurel is possibly the easiest herb to grow. As long as you keep it alive while it is establishing roots (don't let it dry out while rooting), it will provide you with fresh herbs for as long as you live in your house.

You can grow bay laurel in a pot. Just be careful not to overwater it.

Bay laurel is a small tree or large shrub.

CHIVES *(Allium schoenoprasum)*

Chives are a perennial herb that sometimes remains evergreen throughout the winter. This robust plant forms large clumps that can be divided and shared with friends. You can use chives in pastas, fresh salsas, and roasted or grilled dishes.

Both the flowers and the stems of chives are edible.

▪ Recommended Varieties

There aren't really different varieties of chives. Plant whatever you can find at the garden center.

▪ When and Where to Plant

Temperature: Plant chives outside in spring when soil temperatures are at least 60 degrees F.

Soil: Chives are not picky about soil; they grow best in well-draining soils.

Sun: Full sun

▪ How to Plant

Starting seeds indoors: Start seeds indoors one month before planting outside.

Planting outside: Chives are easy to grow from seed. Sow seeds outdoors and cover with seed-starting mix. Keep seeds moist while germinating.

▪ How to Grow

Water: Keep the soil evenly moist. Give extra water after cutting back.

Fertilizer: No extra fertilizer is needed.

Pest control: Chives have no pest problems.

▪ When and How to Harvest

Keep chives producing leaves by cutting sections of the plant back to 3 inches tall once the whole plant is 6 to 8 inches tall. (Don't cut the entire plant back all the way to the ground.)

Chives are one of the earliest herbs to appear in the garden in spring.

GROWING TIPS

You can let chives flower, or you can cut the flowers off if you want the plants to keep producing leaves. The flowers are edible.

MARJORAM *(Origanum majorana)*

Marjoram is a perennial, semiwoody herb in coastal areas of the Carolinas. Add marjoram leaves to chicken dishes, soups, stews, salads, and salad dressings. This versatile herb adds zing to almost any meal, and you're much more likely to use it frequently if you grow your own plants. Make room in the sunny perennial garden for this herb, because you'll want it to stick around.

Recommended Varieties

There are not a lot of different marjoram varieties. For the most success, purchase plants that have been grown outdoors locally and are well adapted to local soils and temperatures.

When and Where to Plant

Temperature: Transplant outside when all danger of frost has passed.

Soil: Marjoram is well adapted to grow in most any soil, but does best in well-drained soil.

Sun: Full sun

Marjoram is often confused with oregano, but marjoram has a gentler, sweeter flavor.

Let some of the marjoram plants flower to attract pollinators to the vegetable garden.

■ *How to Plant*

Starting seeds indoors: You can start seeds indoors or sow directly outside. It is easier to just grow from transplants.

Planting outside: Space plants 8 inches apart. You really only need one or two marjoram plants, and they can be worked into the flower garden.

■ *How to Grow*

Water: Water while the plant is getting established. Avoid overwatering this moisture-sensitive plant.

Fertilizer: Feed once a year in spring.

Pest control: Spider mites can attack marjoram if there's a severe drought in the area and plants get stressed. You can prevent these pest problems by watering plants during dry times.

■ *When and How to Harvest*

Snip off the tender new growth for use in fresh salads. You can use woodier, older growth in soups and roasts.

GROWING TIP

To keep plants producing leaves, snip off the flowers before they open.

Grow mint in a pot to prevent it from taking over the garden.

MINT *(Mentha* spp.*)*

There are two types of perennial mint commonly grown in the Carolinas: peppermint (*Mentha* × *piperita*) and spearmint (*Mentha spicata*). Both types are so prolific that they're practically weeds. If you want to grow mint, I'd advise growing it in a pot where it can't escape and take over the entire garden.

■ *Recommended Varieties*

'Kentucky Colonel' is a spearmint and the official mint of the Kentucky Derby. It tastes great in mint juleps. 'Chocolate' mint really does taste like a YORK Peppermint Pattie. There are many different varieties of mint. Experiment!

When and Where to Plant

Temperature: Mints are semievergreen but will become established best if planted after danger of frost has passed.

Soil: Mint is widely adaptable to all soils.

Sun: Full sun to partial shade

How to Plant

Starting seeds indoors: You don't really grow mint from seed. Grow it from cuttings. If you know anyone who has mint, just cut a few stems and stick them in water to root.

Planting outside: Once the cuttings have rooted, plant them outside in a large pot.

> **GROWING TIP**
>
> It bears repeating: mint can be highly invasive. Grow it in a pot.

How to Grow

Water: Mint is a somewhat heavy drinker. Keep plants moist.

Fertilizer: No extra fertilizer is needed.

Pest control: Mint has no pest problems.

When and How to Harvest

Cut off the tops of the plants to use. Stick in a glass of water until you need to use the leaves.

Children will appreciate the many different varieties of mint.

OREGANO *(Origanum vulgare* subsp. *hirtum)*

Oregano is a perennial herb that's so easy to grow, it's almost a weed. Thankfully, though part of the mint family, it doesn't spread as crazily as mint. Oregano grows in large clumps. It is semievergreen in the Carolinas.

■ *Recommended Varieties*

There are a few different varieties of oregano, but not enough to recommend one over the other.

■ *When and Where to Plant*

Temperature: Plant oregano outside after danger of frost has passed.
Soil: Oregano grows best in soil with a pH of 6.5 to 7.5
Sun: Full sun

■ *How to Plant*

Starting seeds indoors: You can start seeds indoors. Sprinkle them on top of the soil, water them, and keep the seeds moist as the plants germinate.

Planting outside: You can also sow seeds outside directly on top of the soil and use a piece of plastic wrap to keep them moist. However, it is just as easy to buy a couple of plants from the garden center and call it a day. You definitely don't need more than one or two.

Oregano is often known as the "pizza" herb.

Growing oregano in a pot ensures it won't hog all of the space in your vegetable garden.

How to Grow

Water: No extra water is needed after plants have established themselves.

Fertilizer: No extra fertilizer is needed.

Pest control: Spider mites may attack oregano when it's exceptionally dry outside.

When and How to Harvest

Leaves are most fragrant just before the flowers open, but you can snip leaves off to use at any time.

GROWING TIP

The primary objective with oregano is to keep it from taking over the garden. Even though it is a clumping perennial herb, the stems will form roots and spread if allowed to. Periodically cut the entire plant back. It will grow out amazingly fast.

ROSEMARY *(Rosmarinus officinalis)*

Rosemary is a woody, evergreen perennial herb that grows to be a medium-sized shrub in most of the Carolinas (hardy to zone 7). It is native to the Mediterranean and thrives in well-drained soil. It's easy to grow once established and so fun to cook with. You can use fresh rosemary when roasting meats and vegetables. Rosemary twigs make excellent kebab sticks for grilling. Save a corner of the garden for this tasty herb.

■ Recommended Varieties

Rosemary is hardy in all but the coldest parts of our area. There are many neat varieties to try. 'Tuscan Blue' rosemary has an upright growth form with long, straight stems that are great for making shish kebabs for grilling. 'Arp' rosemary is one of the most cold-hardy varieties. 'Benenden Blue' has dark blue flowers. 'Prostrata' is a creeping variety that is less hardy and has light blue flowers. 'Blue Spires' rosemary is a tall, spreading variety that is cold hardy to zone 6.

■ When and Where to Plant

Temperature: Plant transplants outside after danger of frost has passed, for best establishment.

Soil: Rosemary is widely adaptable to many different soil types.

Sun: Full sun

■ How to Plant

Starting seeds indoors: Not recommended

Planting outside: It is much easier to grow rosemary from transplants than to attempt to grow it from seed. Plant transplants outside where they'll have room to grow to be at least 3 feet wide and 3 feet tall over time.

GROWING TIPS

Leave rosemary alone. If you baby it, you're more likely to kill it. Once it is established, don't water it unless your area is experiencing a severe drought. Don't feed, as that promotes weak growth that's susceptible to pests. Rosemary is extremely deer resistant.

■ *When and How to Grow*

Water: No extra water is needed once plants are established.

Fertilizer: No extra fertilizer is needed.

Pest control: Spider mites can occasionally afflict rosemary. Use horticultural oil to treat the plant for these pests.

■ *When and How to Harvest*

Cut pieces of rosemary at any time of the year to use for cooking. The young tips are best for mixing into fresh salads. The rest of the plant is good for use in soups, stews, or roasting.

Rosemary is equally prized for its beautiful winter flowers as it is for its fragrant leaves.

Sage grows best from cuttings.

SAGE *(Salvia officinalis)*

The sage that we use in cooking is called "culinary sage," and all varieties are from the *officinalis* species of salvia. This tough perennial is extremely hardy but needs well-drained soil in order to grow. In the Carolinas, because perennial garden sage doesn't die back to the ground every year, the stems can become woody, and the plant can start producing fewer fragrant leaves. If you have an old sage plant, cut it back to spur on new growth.

▉ *Recommended Varieties*

Beyond the straight species of garden sage, there are many varieties with interesting colors of leaves, increased cold hardiness, and other desirable characteristics. 'Berggarten' sage is a hardier, more compact variety than the straight species of sage. 'Purpurascens' is also a garden sage that has deep purple leaves. 'Tricolor' sage has leaves that are purple, green, and white. It is decorative and useful. 'Icterina' sage has variegated green and yellow leaves.

GROWING TIPS

For more fragrant leaves, cut off sage flowers when they appear. The plant will channel more of its energy into leaf production, which is the part you cook with anyway.

Replant garden sage every three to four years for fresh, new growth. (You can take cuttings off your existing plants for replanting.)

When and Where to Plant

Temperature: Garden sage is hardy in our area, but it produces the most new growth in summer when temperatures are warm. While the leaves might not fall off the plant in winter, the plant does not produce new leaves in that season (except in zone 9b or warmer).

Soil: Plant in well-drained soil.

Sun: Garden sage needs full sun to be most productive and for leaves to be their most fragrant.

How to Plant

Starting seeds indoors: Sage is easier to grow from cuttings than from seed. To take cuttings, use pruners or snips to cut off the top 4 inches of tender new growth. Strip off leaves from the lower half and stick it in potting soil. Once the plant has established good roots (you can pull up on the plant without uprooting it), transplant it outside.

Planting outside: Plant outside when all danger of frost has passed, for best establishment.

How to Grow

Water: Once established, sage needs no extra water. In fact, extra water can actually harm the plant.

Fertilizer: Sage does not usually need extra fertilizer, but a once-a-year feeding in spring with a slow-release, balanced fertilizer won't hurt the plant.

Pest control: Slugs may attack sage when conditions are wetter than usual, and spider mites can be a problem during times of extreme drought. Both problems will go away on their own, eventually.

When and How to Harvest

Cut stems or individual leaves to use when cooking.

TARRAGON *(Artemisia dracunculus)*

Tarragon is another herb that you're much more likely to use if you grow it. I always put fistfuls of leaves in chicken noodle soup. The fragrant, tasty herb really jazzes up an otherwise bland dish. This perennial herb is hardy in the Carolinas and semievergreen. You can harvest fresh leaves year-round.

Recommended Varieties

French tarragon is the more flavorful type, but it struggles in areas with long, hot summers. If you have trouble growing this plant, try Mexican tarragon (*Tagetes lucida*).

When and Where to Plant

Temperature: Plant tarragon transplants outside after danger of frost has passed.

Soil: Tarragon grows best in soil with a pH of 6.0 to 7.0.

Sun: Full sun to partial shade

Tarragon can be harvested year round.

How to Plant

Starting seeds indoors: You can't grow French tarragon from seed, only from cuttings. Take cuttings in early spring and root them before planting outside.

Planting outside: Plant outside, allowing 12 to 18 inches between plants.

How to Grow

Water: Tarragon requires an average amount of water. Water tarragon when you water your annual flowers.

Fertilizer: No fertilizer is necessary.

Pest control: There are no pests that regularly attack tarragon.

When and How to Harvest

Cut leaves to use in cooking, leaving at least 6 inches on the plant when cutting branches to use.

GROWING TIP

Plant tarragon in the vegetable garden to keep pests away from other plants.

Try Mexican tarragon in areas with long, hot summers.

Fresh or dried, thyme makes a great flavoring for meats or vinegars.

THYME *(Thymus vulgaris)*

Thyme is a perennial herb that is hardy in zones 5–9. There are over a hundred varieties of thyme, but only a few are proven performers in the Carolinas. Thyme is indispensible for cooking savory dishes. Use it in everything from roast chicken to vegetable soup to shortbread cookies. If you're interested in making your own flavored vinegars, thyme is a good herb to try. Thyme can be cut back or replanted every few years for fresh new growth.

Recommended Varieties

Lemon thyme and lime thyme (*Thymus × citriodorus*) grow well in the Carolinas. 'English' thyme and 'French' thyme are commonly available varieties. 'Doone Valley' is a creeping variety with leaves that turn a yellow/green variegated pattern during cooler weather. There are many varieties of thyme—try several in your garden, and see which ones you like best.

When and Where to Plant

Temperature: Thyme grows best in warm temperatures. Plant thyme plants outside after all danger of frost has passed, for best results.

Soil: Thyme needs well-drained soil to grow. It will rot if kept wet.

Sun: Full sun

GROWING TIPS

If you have heavy clay soil that drains poorly, grow thyme in a pot. Plant thyme in the garden along with other vegetables to help repel pests.

How to Plant

Starting seeds indoors: Not recommended

Planting outside: It is much easier to grow thyme from cuttings than it is to grow it from seed. Root cuttings by snipping off 4-inch pieces of fresh, green growth from a plant. Strip the leaves off the lower half and place it in a cup of water. Once the plant has developed roots, plant it in a 4-inch pot. When the plant has rooted in, transplant it outside.

How to Grow

Water: Thyme does not need extra water once established.

Fertilizer: Fertilize once a year in spring with a balanced, slow-release fertilizer.

Pest control: Spider mites will attack the plant if it is stressed—usually during drought. Fungal diseases and root rot are also problems if the soil is too wet.

When and How to Harvest

Thyme leaves are most fragrant if they are cut before the plant flowers. Keep cutting back the top 4 to 6 inches of growth to stimulate fresh, new growth, which is best for cooking.

Harvest thyme before the flowers appear.

RESOURCES

This list is not comprehensive, but it includes numerous sources to get you going in your search for more information, plant suppliers, and local seed growers.

Nurseries and Plant Suppliers

Seeds

Southern Exposure Seed Exchange
(seeds, sweet potato slips, garlic bulbs)
http://www.southernexposure.com/
PO Box 460
Mineral, VA 23117
E-mail: gardens@southernexposure.com
Phone: 540-894-9480
Fax: 540-894-9481

Seeds for the South
(untreated heirloom vegetable, herb, tomato, and pepper seeds
http://www.seedsforthesouth.com/
Vegetable Seed Warehouse
P.O. Box 13741
Charleston, SC 29422
E-mail: Orders@vegetableseedwarehouse.com

Penny's Tomatoes
http://www.pennystomatoes.com/
3904 Kensington Ct.
Myrtle Beach, SC 29577
Email: penny@pennystomatoes.com
Phone: 843-626-4507
Fax: 843-626-4507

Pepper Joe's, Inc.
http://www.pepperjoe.com/
725 Carolina Farm Blvd.
Myrtle Beach, SC 29579
Email: pepperjoe1@sc.rr.com
Phone: 843-742-5116

Sow True Seed
(herbs, vegetables, flowers, bulbs and tubers, cover crops and grains)
http://sowtrueseed.com
146 Church Street
Asheville, NC 28801
Email: info@sowtrue.com
Phone: 828-254-0708
Fax: 828-254-0708

Citrus

McKenzie Farms
http://mckenzie-farms.com/index.htm
2115 Olanta Highway
Scranton, SC 29591
Email: citrusman99@hotmail.com
Phone: 843-389-4831

Apples

Big Horse Creek Farm
(antique and heirloom apple trees)
http://www.bighorsecreekfarm.com
PO Box 70
Lansing, NC 28643
E-mail: oldapple@bighorsecreekfarm.com

Blueberries

Finch Blueberry Nursery
http://www.danfinch.com/berrys.htm
PO Box 699
Bailey, NC 27807
Email: finchnursery@bbnp.com
Phone: 800-245-4662 (toll free), 252-235-4664
Fax: 252-235-2411

Fruit Trees

Johnson Nursery
http://www.johnsonnursery.com/
1352 Big Creek Road
Ellijay, GA 30536
E-mail: sales@johnsonnursery.com
Phone: 888-276-3187 (toll-free); 706-276-3187
Fax: 706-276-3186

Stark Bro's
http://www.starkbros.com/
11523 Highway NN
Louisiana, MO 63353
While Stark Bro's ships around the United States, they have a quality product and get good feedback from customers.
Email: info@starkbros.com
Phone: 573-754-3113

Websites

Clemson University Plant Problem Clinic and Nematode Assay Lab
http://www.clemson.edu/public/regulatory/plant_industry/plant_prob_clinic/
This site includes information about how to submit plant samples for problem testing and identification.

Dave's Garden
http://davesgarden.com/
One of the longest-running websites, forums, and plant swap groups online. There's a group of gardeners for every single plant interest and question. There's also a robust rating system for mail-order nurseries.

Frost Tables
http://cdo.ncdc.noaa.gov/cgi-bin/climatenormals/climatenormals.pl?directive=prod_
select2&prodtype=CLIM2001&subrnum
Select your state to find the first and last frost dates for specific cities. Not all cities are listed.

North Carolina Lawn and Garden
(Cooperative Extension at North Carolina State University)
http://www.ces.ncsu.edu/categories/lawn-garden/

North Carolina State University Plant Disease and Insect Clinic
http://ncsupdicblog.blogspot.com/
This is a blog devoted to plant pests and diseases in North Carolina.

South Carolina Home and Garden Information Center
(Cooperative Extension at Clemson University)
http://www.clemson.edu/extension/hgic/

USDA Plant Hardiness Zone Map
http://planthardiness.ars.usda.gov/PHZMWeb/
Click on your state or enter your zip code and locate the hardiness zone for your individual area.

Books

Beginner's Illustrated Guide to Gardening: Techniques to Help You Get Started by
Katie Elzer-Peters
If you're a completely new gardener, pick up my other book, which is filled with step-by-step instructions for many of the most important gardening tasks. (It covers vegetable gardening and landscape maintenance.) Starting with a good foundation will help you be a more successful gardener.

GLOSSARY

Acidic soil: On a soil pH scale of 0 to 14, acidic soil has a pH lower than 5.5. Most garden plants prefer a soil a bit on the acidic side.

Afternoon sun: A garden receiving afternoon sun typically has full sun from 1:00 to 5:00 p.m. daily, with more shade during the morning hours.

Alkaline soil: On a soil pH scale of 0 to 14, alkaline soil has a pH higher than 7.0. Many desert plants thrive in slightly alkaline soils.

Annual: A plant that germinates (sprouts), flowers, and dies within one year or season (spring, summer, winter, or fall) is an annual.

***Bacillus thuringiensis* (B.t.):** *B.t.* is an organic pest control based on naturally occurring soil bacteria, often used to control harmful caterpillars such as cutworms, leaf rollers, and webworms.

Balled and burlapped (B&B): This phrase describes plants that have been grown in field nursery rows, dug up with their soil intact, wrapped with burlap, and tied with twine. Most of the plants sold balled and burlapped are large evergreen plants and deciduous trees.

Bare root: Bare-root plants are those that are shipped dormant, without being planted in soil or having soil around their roots. Roses are often shipped bare root.

Beneficial insects: These insects perform valuable services such as pollination and pest control. Ladybugs, soldier beetles, and some bees are examples.

Biennial: A plant that blooms during its second year and then dies is a biennial.

Bolting: This is a process when a plant switches from leaf growth to producing flowers and seeds. Bolting often occurs quite suddenly and is usually undesirable, because the plant usually dies shortly after bolting.

Brown materials: A part of a well-balanced compost pile, brown materials include high-carbon materials such as brown leaves and grass, woody plant stems, dryer lint, and sawdust.

Bud: The bud is an undeveloped shoot nestled between the leaf and the stem that will eventually produce a flower or plant branch.

Bulb: A bulb is a plant with a large, rounded underground storage organ formed by the plant stem and leaves. Examples are tulips, daffodils, and hyacinths. Bulbs that flower in spring are typically planted in fall.

Bush: *See* shrub.

Cane: A stem on a fruit shrub; usually blackberry or raspberry stems are called canes, but blueberry stems can also be referred to as canes.

Central leader: The term for the center trunk of a fruit tree.

Chilling hours: Hours when the air temperature is below 45 degrees F; chilling hours are related to fruit production.

Common name: A name that is generally used to identify a plant in a particular regions, as opposed to its botanical name, which is standard throughout the world; for example, the common name for *Echinacea purpurea* is "purple coneflower."

Contact herbicide: This type of herbicide kills only the part of the plant that it touches, such as the leaves or the stems.

Container: Any pot or vessel that is used for planting; containers can be ceramic, clay, steel, or plastic—or a teacup, bucket, or barrel.

Container garden: This describes a garden that is created primarily by growing plants in containers instead of in the ground.

Container grown: This describes a plant that is grown, sold, and shipped while in a pot.

Cool-season annual: This is a flowering plant, such as snapdragon or pansy, that thrives during cooler months.

Cool-season vegetable: This is a vegetable, such as spinach, broccoli, and peas, that thrives during cooler months.

Cover crop: These plants are grown specifically to enrich the soil, prevent erosion, suppress weeds, and control pests and diseases.

Cross-pollinate: This describes the transfer of pollen from one plant to another plant.

Dappled shade: This is bright shade created by high tree branches or tree foliage, where patches of sunlight and shade intermingle.

Day-neutral plant: A plant that flowers when it reaches a certain size, regardless of the day length, is a day-neutral plant.

Deadhead: To remove dead flowers in order to encourage further bloom and prevent the plant from going to seed is to deadhead.

Deciduous plant: A plant that loses its leaves seasonally, typically in fall or early winter, is deciduous.

Diatomaceous earth: A natural control for snails, slugs, flea beetles, and other garden pests, diatomaceous earth consists of ground-up fossilized remains of sea creatures.

Dormancy: The period when plants stop growing in order to conserve energy, this happens naturally and seasonally, usually in winter.

Drip line: The ground area under the outer circumference of tree branches, this is where most of the tree's roots that absorb water and nutrients are found.

Dwarf: In the context of fruit gardening, a dwarf fruit tree is a tree that grows no taller than 10 feet tall and is usually a dwarf as a result of the rootstock of the tree.

Evergreen: A plant that keeps its leaves year-round, instead of dropping them seasonally is evergreen.

Floricane: A second-year cane on a blackberry or raspberry shrub; floricanes are fruit bearing.

Flower stalk: The stem that supports the flower and elevates it so that insects can reach the flower and pollinate it is the flower stalk.

Four-tine claw: Also called a cultivator, this hand tool typically has three to four curved tines and is used to break up soil clods or lumps before planting and to rake soil amendments into garden beds.

Frost: Ice crystals that form when the temperature falls below freezing (32 degrees F) create frost.

Full sun: Areas of the garden that receive direct sunlight for six to eight hours a day or more, with no shade, are in full sun.

Fungicide: This describes a chemical compound used to control fungal diseases.

Gallon container: A standard nursery-sized container for plants, a gallon container is roughly equivalent to a gallon container of milk.

Garden fork: A garden implement with a long handle and short tines, use a garden fork for loosening and turning soil.

Garden lime: This soil amendment lowers soil acidity and raises the pH.

Garden soil: The existing soil in a garden bed; it is generally evaluated by its nutrient content and texture. Garden soil is also sold as a bagged item at garden centers and home-improvement stores.

Germination: This is the process by which a plant emerges from a seed or a spore.

Grafted tree: This is a tree composed of two parts: the top, or scion, which bears fruit, and the bottom, or rootstock.

Graft union: This is the place on a fruit tree trunk where the rootstock and the scion have been joined.

Granular fertilizer: This type of fertilizer comes in a dry, pellet-like form rather than a liquid or powder.

Grass clippings: The parts of grass that are removed when mowing, clippings are a valuable source of nitrogen for the lawn or the compost pile.

Green materials: An essential element in composting that includes grass clippings, kitchen scraps, and manure and provides valuable nitrogen in the pile, green materials are high in nitrogen.

Hand pruners: An important hand tool that consists of two sharp blades that perform a scissoring motion, these are used for light pruning, clipping, and cutting.

Hardening off: This is the process of slowly acclimating seedlings and young plants grown in an indoor environment to the outdoors.

Hardiness zone map: This map lists average annual minimum temperature ranges of a particular area. This information is helpful in determining appropriate plants for the garden. North America is divided into eleven separate hardiness zones.

Hard rake: This tool has a long handle and rigid tines at the bottom. It is great for moving a variety of garden debris, such as soil, mulch, leaves, and pebbles.

Hedging: This is the practice of trimming a line of plants to create a solid mass for privacy or garden definition.

Heirloom: A plant that was more commonly grown during earlier periods of human history but is not widely used in modern commercial agriculture is an heirloom plant.

Hoe: A long-handled garden tool with a short, narrow, flat steel blade, it is used for breaking up hard soil and removing weeds.

Hose breaker: This device screws onto the end of a garden hose to disperse the flow of water pressure from the hose.

Host plant: A plant grown to feed caterpillars that will eventually morph into butterflies is called a host plant.

Hybrid: Plants produced by crossing two genetically different plants, hybrids often have desirable characteristics such as disease resistance.

Insecticide: This substance is used for destroying or controlling insects that are harmful to plants. Insecticides are available in organic and synthetic forms.

Irrigation: A system of watering the landscape, irrigation can be an in-ground automatic system, soaker or drip hoses, or hand-held hoses with nozzles.

Jute twine: A natural-fiber twine, jute is used for gently staking plants or tying them to plant supports.

Larva: The immature stage of an insect that goes through complete metamorphosis; caterpillars are butterfly or moth larvae.

Larvae: This is the plural of larva.

Leaf rake: A long-handled rake with flexible tines on the head, a leaf rake is used for easily and efficiently raking leaves into piles.

Liquid fertilizer: Plant fertilizer in a liquid form, some types need to be mixed with water, and some types are ready to use from the bottle.

Long-day plant: Plants that flower when the days are longer than their critical photoperiod, long-day plants typically flower in early summer, when the days are still getting longer.

Loppers: One of the largest manual gardening tools, use loppers for pruning branches of 1 to 3 inches in diameter with a scissoring motion.

Morning sun: Areas of the garden that have an eastern exposure and receive direct sun in the morning hours are in morning sun.

Mulch: Any type of material that is spread over the soil surface around the base of plants to suppress weeds and retain soil moisture is mulch.

Nematode: Microscopic, wormlike organisms that live in the soil, some nematodes are beneficial, while others are harmful.

New wood (new growth): The new growth on plants, it is characterized by a greener, more tender form than older, woodier growth.

Nozzle: A device that attaches to the end of a hose and disperses water through a number of small holes; the resulting spray covers a wider area.

Old wood: Old wood is growth that is more than one year old. Some fruit plants produce on old wood. If you prune these plants in spring before they flower and fruit, you will cut off the wood that will produce fruit.

Organic: This term describes products derived from naturally occurring materials instead of materials synthesized in a lab.

Part shade: Areas of the garden that receive three to six hours of sun a day are in part shade. Plants requiring part shade will often require protection from the more intense afternoon sun, either from tree leaves or from a building.

Part sun: Areas of the garden that receive three to six hours of sun a day are in part sun. Although the term is often used interchangeably with "part shade," a "part sun" designation places greater emphasis on the minimal sun requirements.

Perennial: A plant that lives for more than two years is a perennial. Examples include trees, shrubs, and some flowering plants.

pH: A figure designating the acidity or the alkalinity of garden soil, pH is measured on a scale of 1 to 14, with 7.0 being neutral.

Pinch: This is a method to remove unwanted plant growth with your fingers, promoting bushier growth and increased blooming.

Pitchfork: A hand tool with a long handle and sharp metal prongs, a pitchfork is typically used for moving loose material such as mulch or hay.

Plant label: This label or sticker on a plant container provides a description of the plant and information on its care and growth habits.

Pollination: The transfer of pollen for fertilization from the male pollen-bearing structure (stamen) to the female structure (pistil), usually by wind, bees, butterflies, moths, or hummingbirds; this process is required for fruit production.

Potting soil: A mixture used to grow flowers, herbs, and vegetables in containers, potting soil provides proper drainage and extra nutrients for healthy growth.

Powdery mildew: A fungal disease characterized by white powdery spots on plant leaves and stems, this disease is worse during times of drought or when plants have poor air circulation.

Power edger: This electric or gasoline-powered edger removes grass along flower beds and walkways for a neat appearance.

Pre-emergent herbicide: This weedkiller works by preventing weed seeds from sprouting.

Primocane: A first-year cane on a blackberry shrub, a primocane doesn't produce fruit.

Pruning: This is a garden task in which a variety of hand tools are used to remove dead or overgrown branches to increase plant fullness and health.

Pruning saw: This hand tool for pruning smaller branches and limbs features a long, serrated blade with an elongated handle.

Rhizome: An underground horizontal stem that grows side shoots, a rhizome is similar to a bulb.

Rootball: The network of roots and soil clinging to a plant when it is lifted out of the ground is the rootball.

Rootstock: The bottom part of a grafted fruit tree, rootstocks are often used to create dwarf fruit trees, impart pest or disease resistance, or make a plant more cold hardy.

Runner: A stem sprouting from the center of a strawberry plant, a runner produces fruit in its second year.

Scaffold branch: This horizontal branch emerges almost perpendicular to the trunk.

Scientific name: This two-word identification system consists of the genus and species of a plant, such as *Ilex opaca*.

Scion: The top, fruit-bearing part of a grafted fruit tree is the scion.

Scissors: A two-bladed hand tool great for cutting cloth, paper, twine, and other lightweight materials, scissors are a basic garden tool.

Seed packet: The package in which vegetable and flower seeds are sold, it typically includes growing instructions, a planting chart, and harvesting information.

Seed-starting mix: Typically a soilless blend of perlite, vermiculite, peat moss, and other ingredients, seed-starting mix is specifically formulated for growing plants from seed.

Self-fertile: A plant that does not require cross-pollination from another plant in order to produce fruit is self-fertile.

Semidwarf: A fruit tree grafted onto a rootstock that restricts growth of the tree to one-half to two-thirds of its natural size is semidwarf.

Shade: Garden shade is the absence of any direct sunlight in a given area, usually due to tree foliage or building shadows.

Short-day plant: Flowering when the length of day is shorter than its critical photoperiod, short-day plants typically bloom during fall, winter, or early spring.

Shovel: A handled tool with a broad, flat blade and slightly upturned sides, used for moving soil and other garden materials, a shovel is a basic garden tool.

Shredded hardwood mulch: A mulch consisting of shredded wood that interlocks, resisting washout and suppressing weeds, hardwood mulch can change soil pH.

Shrub: This woody plant is distinguished from a tree by its multiple trunks and branches and its shorter height of less than 15 feet tall.

Shrub rake: This long-handled rake with a narrow head fits easily into tight spaces between plants.

Sidedress: To sprinkle slow-release fertilizer along the side of a plant row or plant stem is to sidedress.

Slow-release fertilizer: This form of fertilizer releases nutrients at a slower rate throughout the season, requiring less-frequent applications.

Snips: This hand tool, used for snipping small plants and flowers, is perfect for harvesting fruits, vegetables, and flowers.

Soaker hose: This is an efficient watering system in which a porous hose, usually made from recycled rubber, allows water to seep out around plant roots.

Soil knife: This garden knife with a sharp, serrated edge, is used for cutting twine, plant roots, turf, and other garden materials.

Soil test: An analysis of a soil sample, this determines the level of nutrients (to identify deficiencies) and detects pH.

Spade: This short-handled tool with a sharp, rectangular metal blade is used for cutting and digging soil or turf.

Spur: This is a small, compressed, fruit-bearing branch on a fruit tree.

Standard: Describing a fruit tree grown on its own seedling rootstock or a nondwarfing rootstock, this is the largest of the three sizes of fruit trees.

Sucker: The odd growth from the base of a tree or a woody plant, often caused by stress, this also refers to sprouts from below the graft of a rose or fruit tree. Suckers divert energy away from the desirable tree growth and should be removed.

Systemic herbicide: This type of weedkiller is absorbed by the plant's roots and taken into the roots to destroy all parts of the plant.

Taproot: This is an enlarged, tapered plant root that grows vertically downward.

Thinning: This is the practice of removing excess vegetables (root crops) to leave more room for the remaining vegetables to grow; also refers to the practice of removing fruits when still small from fruit trees so that the remaining fruits can grow larger.

Topdress: To spread fertilizer on top of the soil (usually around fruit trees or vegetables) is to topdress.

Transplants: Plants that are grown in one location and then moved to and replanted in another, seeds started indoors and nursery plants are two examples.

Tree: This woody perennial plant typically consists of a single trunk with multiple lateral branches.

Tree canopy: This is the upper layer of growth, consisting of the tree's branches and leaves.

Tropical plant: This is a plant that is native to a tropical region of the world, and thus acclimated to a warm, humid climate and not hardy to frost.

Trowel: This shovel-like hand tool is used for digging or moving small amounts of soil.

Warm-season vegetable: This is a vegetable that thrives during the warmer months. Examples are tomatoes, okra, and peppers. These vegetables do not tolerate frost.

Watering wand: This hose attachment features a longer handle for watering plants beyond reach.

Water sprout: This vertical shoot emerges from a scaffold branch. It is usually nonfruiting and undesirable.

Wheat straw: These dry stalks of wheat, which are used for mulch, retain soil moisture and suppress weeds.

Wood chips: Small pieces of wood made by cutting or chipping, wood chips are used as mulch in the garden.

INDEX

PHOTO CREDITS

MEET KATIE ELZER-PETERS

Gardening is a hobby-turned-career for Katie Elzer-Peters, one her parents and grandparents nurtured in her since she could walk. After receiving a bachelor's of science in public horticulture from Purdue University, Katie completed the Longwood Graduate Program at Longwood Gardens and the University of Delaware, receiving a master's of science in public garden management.

Katie has served as a horticulturist, head of gardens, educational programs director, development officer, and manager of botanical gardens around the United States, including the Washington Park Arboretum in Seattle, Washington; the Indianapolis Zoo in Indianapolis, Indiana; the Marie Selby Botanical Garden in Sarasota, Florida; the Smithsonian Institution in Washington, DC; Longwood Gardens in Kennett Square, Pennsylvania; Winterthur Museum, Garden, and Library in Greenville, Delaware; the King's Garden at Fort Ticonderoga in Ticonderoga, New York; and Airlie Gardens in Wilmington, North Carolina.

Whether at a botanical garden, or for a garden center, garden club, or school group, Katie has shared her love of gardening by teaching classes and workshops and writing brochures, articles, gardening website information, and columns. While serving as curator of landscape at Fort Ticonderoga, Katie planned and led garden bus tours along the east coast of the United States and Canada.

Today, Katie lives and gardens with her husband and their dog in the coastal city of Wilmington, North Carolina (zone 8a). She loves the year-round weather for gardening but is in constant battle with the sandy soil. Katie manages GreatGardenSpeakers.com, an online speaker directory of garden, design, ecology, and horticultural speakers. She also owns the Garden of Words, LLC, a marketing and PR firm specializing in garden-industry clients.

Katie is the author of *Beginner's Illustrated Guide to Gardening: Techniques to Help You Get Started*, also published by Cool Springs Press.

NOTES